SEPTEMBER 2021

CLIMATE-RESILIENCE POLICIES AND INVESTMENTS FOR EGYPT'S AGRICULTURE SECTOR

SUSTAINING PRODUCTIVITY AND FOOD SECURITY

NICOSTRATO PEREZ, YUMNA KASSIM, CLAUDIA RINGLER, TIMOTHY S. THOMAS, HAGAR ELDIDI, AND CLEMENS BREISINGER

A Peer-Reviewed Publication
International Food Policy Research Institute
Washington, DC

ABOUT IFPRI

The International Food Policy Research Institute (IFPRI), a CGIAR center established in 1975, provides research-based policy solutions to sustainably reduce poverty and end hunger and malnutrition. IFPRI's strategic research aims to foster a climate-resilient and sustainable food supply; promote healthy diets and nutrition for all; build inclusive and efficient markets, trade systems, and food industries; transform agricultural and rural economies; and strengthen institutions and governance. Gender is integrated in all the Institute's work. Partnerships, communications, capacity strengthening, and data and knowledge management are essential components to translate IFPRI's research from action to impact. The Institute's regional and country programs play a critical role in responding to demand for food policy research and in delivering holistic support for country-led development. IFPRI collaborates with partners around the world.

Photo credits: (cover) Jeremy Bezanger/Unsplash

ISBN: 978-0-89629-418-9

DOI: https://doi.org/10.2499/9780896294189

CONTENTS

ACKNOWLEDGMENTS 6

EXECUTIVE SUMMARY 7

INTRODUCTION 10

1. THE AGRICULTURE SECTOR 11

2. CLIMATE CHANGE AND RESILIENCY OF THE AGRICULTURE SECTOR 19

3. DEVELOPING RESILIENCE AGAINST CLIMATE CHANGE 27

4. CONCLUSIONS AND POLICY IMPLICATIONS: INVESTMENTS FOR SUSTAINABLE AND RESILIENT AGRICULTURE 39

REFERENCES 43

APPENDIX 45

ABOUT THE AUTHORS 51

TABLES

TABLE 1.1 Land-use area and change in Egypt, 1990-2020 12

TABLE 1.2 Sources and uses of water resources in Egypt, crop years 2012/13 to 2015/16 12

TABLE 1.3 Agriculture and the Egyptian economy, 1990, 2000, 2010, 2015-2019 14

TABLE 1.4 Recent performance of Egypt's agriculture sector – food production growth, 1990-2020 16

TABLE 1.5 Trade performance of Egyptian agriculture, 1990-2020 17

TABLE 2.1 Seasonal and annual climate for Egyptian agricultural areas, 1970-2000, and changes through 2040-2070 for 3 climate scenarios from CMIP5 20

TABLE 2.2 Changes in productivity due to biophysical and economic effects of climate change, Egypt, projected by 2050 22

TABLE 2.3 Projected impact of climate change on food production in Egypt and the rest of the world, 2050 24

TABLE 2.4 Impact of climate change on food trade, Egypt, projected by 2050 25

TABLE 2.5 Changes in per capita food and daily calorie availability, Egypt, by 2050 25

TABLE 2.6 Changes in society's welfare due to impacts of climate change on agriculture, world and Egypt, 2020-2050. 26

TABLE 3.1 Yield effects of moderate adoption of suite of climate-resilient technologies for major food groups, Egypt, by 2050 (moderate adoption rate) 32

TABLE 3.2 Production effects of adoption of a suite of climate-resilient technologies for major food groups, Egypt, by 2050 (moderate adoption rate) 34

TABLE 3.3 Changes in net trade status under climate change and adoption of suite of climate-resilient technologies, Egypt, by 2050 (moderate adoption rate, only Egypt adapting) 35

TABLE 3.4 Impact of climate change and adaptation strategies on world prices of agricultural commodities, by 2050 (moderate adoption rate for both Egypt and ROW) 37

TABLE 3.5 Changes in net trade of major food crops due to climate change and adaptation strategies, Egypt and rest of the world, by 2050 (moderate adoption rate) 38

TABLE 3.6 Changes in per capita food intake and calorie consumption due to climate change and adoption of climate-resilient technologies, Egypt, by 2050 (moderate adoption rate for both Egypt and ROW) 38

TABLE A1 Recent performance of Egypt's agriculture sector: Growth of area harvested and animal numbers, 1990-2020 45

TABLE A2 Recent performance of Egypt's agriculture sector: Growth in yield levels, 1990-2020. 46

TABLE A3 Projected rates of yield growth, 2020-2050, under business-as-usual, without climate change, indexed to 2020 47

TABLE A4 Changes in productivity due to climate change: Biophysical and economic effects, rest of the world, by 2050 48

TABLE A5 Effects of climate change on food prices, by 2050. 48

TABLE A6 Yield effects of suite of climate-resilient technologies on major food groups, Egypt, by 2050 (high adoption rate) 49

TABLE A7 Production effects of adoption of suite of climate-resilient technologies on major food groups, Egypt, by 2050 (high adoption rate) 50

TABLE A8 Changes in net trade status under climate change and adoption of suite of climate-resilient technologies, Egypt, by 2050 (high adoption rate) 50

FIGURES

FIGURE 1.1 Satellite image of Egypt: Importance of the Nile River and Delta for agriculture 11

FIGURE 1.2 Trends in production of major food commodities in Egypt, 1990-2020 17

FIGURE 1.3 Production, area harvested, and yield performance of selected food crops in Egypt, 1990-2020 18

FIGURE 2.1 Biophysical and economic modeling framework of the IMPACT model 19

FIGURE 2.2 Projected change in mean daily maximum temperature (°C), 1970-2000 to 2040-2070 21

FIGURE 2.3 Changes in yield trajectories due to combined biophysical and economic effects of climate change for maize, rice, and wheat, projected, 2020-2050. 23

FIGURE 2.4 Projected effects of climate change on global food prices, 2050 25

FIGURE 3.1 Maize, rice, and wheat yields under alternative climate and technology scenarios, Egypt, by 2050 (moderate rate of adoption) 33

FIGURE 3.2 Impact of climate-resilient technologies on production of major food crops, Egypt, by 2050 (moderate adoption rate) 34

FIGURE 3.3 Impact on world prices of major food groups of climate change and suite of climate-resilient technology countermeasures, by 2050 (moderate adoption rate) 37

FIGURE A1 Impact of climate-resilient technologies on production of major food crops, Egypt, by 2050 (high adoption rate) 49

APPENDIXES

A. Additional tables: The agriculture sector and the rest of the world 45

B. Additional figures and tables: High adoption rate 49

ACKNOWLEDGMENTS

We thank Richard Robertson and Hua Xie, both research fellows in IFPRI 's Environment and Production Technology Division (EPTD), for analysis of CMIP5 climate change impacts on Egypt's food crops and for water resource analyses, respectively. We also thank participants, speakers, and commentators at the IFPRI webinar "Beyond COVID-19: Strengthening Climate Resilience in Egypt's Agricultural Sector," held July 21, 2020, at which preliminary results from this study were presented. We thank the contributors to the seminar, including Mohamed El-Kersh, Assistant to the Minister of Agriculture and Land Reclamation; and Yasser ElShayeb, Chief of Party of the Center of Excellence for Water at the American University in Cairo. We also thank Channing Arndt, Director of EPTD at IFPRI for his guidance and support. We appreciate the comments from anonymous reviewers. We gratefully acknowledge the financial support of the United States Agency for International Development (USAID), which made this study possible under the project Evaluating Impact and Building Capacity (EIBC) that is implemented by IFPRI. In addition, we thank Ayat ElDersh and Michael Trueblood for their input and comments to the report and overall guidance and support to the EIBC project.

EXECUTIVE SUMMARY

EGYPT HAS A VIBRANT AND RESILIENT AGRICULTURE SECTOR

Despite its hot desert climate, Egypt has developed a vibrant agriculture sector, chiefly due to large investments in irrigation systems along the Nile River, but also because it has built one of the world's largest agricultural research and development (R&D) systems and supported modern farm mechanization in commercial agriculture.

Agriculture and food production growth overall has been rapid over the last three decades, though for a few commodities, growth has slowed or contracted. In aggregate, meat production grew at an annual rate of 3.8 percent from 1990 to 2020 and cereals at 1.9 percent. Annual production of fruits, vegetables, and roots and tubers (mainly potatoes) for export grew at 3.5, 2.1 and 4.3 percent, respectively.

Egypt's agriculture sector weathered the recent COVID-19 pandemic and economic slowdown relatively well, performing somewhat better than the service and industry sectors. But agriculture is unlikely to do as well under climate change. This report assesses the impacts of climate change on Egypt's agriculture sector, food security, and nutrition; examines alternative adaptation options; and presents policy recommendations aimed at strengthening the Egyptian food system's resilience to climate change.

CLIMATE CHANGE IS A GROWING AND LASTING THREAT TO THE COUNTRY'S AGRICULTURAL SYSTEM

Climate change affects crop productivity globally; in Egypt, large adverse impacts are expected for the country's agriculture and food systems. Direct impacts include higher temperatures and even less precipitation in this already arid country. By 2050, mean daily maximum temperatures for agricultural areas are predicted to increase by 3.1°C and minimum temperatures by 3.4°C over recent levels. Annual rainfall is projected to decline by as much as 15 mm; this is substantial, given that annual rainfall levels average only 42 mm in the country's agricultural areas. At the same time, potential evapotranspiration from crops, an indicator of irrigation water demand, is expected to increase by as much as 21 percent.

We use IFPRI's International Model for Policy Analysis of Agricultural Commodities and Trade (IMPACT) to assess the impact of climate change on Egypt's agriculture and to explore the robustness of different investments in climate change adaptation by comparing the outcomes of these investments to a reference climate change scenario in which no specific adaptation effort is undertaken. In addition, to better understand the differential impact on crop productivity from climate change, a no-climate-change scenario that assumes continuation of historical climate patterns from 1970 to 2000 was added to the analysis. Output from three general circulation models (GCMs) of global climate patterns were used to simulate future climate conditions.

Projected impacts on food production include biophysical impacts from higher temperatures and changes in water availability. They also include economic effects from changes in producer supply and consumer demand, which adjust to climate-induced changes in food prices. Food crop yields in Egypt are projected to decline by 10 percent, on average, as a result of heat stress, water stress, and increased salinity.

Global declines in crop yields and food production caused by climate change are projected to drive global food prices up by as much as 23 percent for maize and 19 percent for rice. Prices of oil crops, root and tuber crops, and poultry are also projected to increase significantly, substantially affecting the food security of poorer segments of Egypt's population. While farmers might be able to increase incomes from commodities they produce, most farmers are net food buyers, and poorer farmers particularly will suffer from price increases. Higher prices, in turn, will reduce Egypt's food import demand, as well as dampening demand for its exports of high-value commodities, such as fruits and vegetables. These tighter food markets will make it more difficult for Egypt to rely on food imports to augment domestic supplies.

IMPACTS OF CLIMATE CHANGE ARE EXACERBATED BY LAND AND WATER SCARCITY

The scarcity of fertile land, which is largely limited to land along the Nile, and the rapid rate of urbanization will further exacerbate the impact of climate change on food production. Around 75,000 ha of fertile land are estimated to have been lost to urban expansion in the last 20 years.

Agriculture is the predominant user of freshwater resources in Egypt, with an estimated consumption level of 62 billion cubic meters (bcm) of water annually, equivalent to 82 percent of total supply. With an increasing population, growing investment in industry, and increasing water loss to evaporation, reduced water availability will be a growing threat to the country's agricultural economy.

THE ECONOMIC COSTS OF CLIMATE CHANGE ARE SUBSTANTIAL

Within the agriculture and food sector alone, the average economic cost, using the three GCMs, of climate change to the Egyptian population is estimated at US$55.3 billion for a 30-year period (2020–2050), or $1.84 billion per year. Consumers bear most of the cost with $2.21 billion per year, while producers gain an average of $0.37 billion per year, implying that higher prices more than compensate for productivity losses. However, most farmers – especially smallholder farmers – are marginal producers and net buyers of food, and thus are expected to suffer net economic losses from the combined producer and consumer effects.

INVESTMENT IN AGRICULTURAL R&D TO ADAPT TO CLIMATE CHANGE WILL PAY OFF

To address adverse climate change impacts, investment in agricultural R&D with a focus on technologies that directly counter climate change effects will be key. Such investment includes both the global research system and national research systems that adapt technologies to local environmental and social conditions. It also involves strengthening extension systems and other information channels for the promotion and dissemination of these technologies.

Rather than simulating a single technology, or "silver bullet," strategy, a "suite of technologies" approach to climate change adaptation was used for this report, assuming farmers tend to combine various resilience strategies to counteract the adverse impacts of climate change. The technology suites considered here include new/improved seeds with traits focused on combating climate change, soil fertility management practices, improved irrigation management, and enhanced crop protection from pests and diseases. As many of these technologies do best when applied in tandem, stacking of complementary technologies and practices was also considered. The five adaptation technology suites were simulated under the with-climate-change scenarios to examine their capacity to reverse some of the adverse impacts of climate change on agricultural productivity in Egypt. Two levels of farm-level adoption of these technology suites were evaluated – a moderate level assuming a 60 percent adoption rate and a high level assuming an 80 percent adoption rate.

For crops less affected by climate change in Egypt, such as fruits and vegetables, potatoes, rice, and wheat, increased investments in climate-change-responsive crop traits or a combination of measures, as well as investments in soil fertility improvement, water management, and crop protection, can counteract the adverse impacts of climate change. Across food crops, investments in climate-resilient seed technologies provide the largest returns, followed by investments in soil fertility, crop protection, and irrigation.

However, for those crops hit hardest by climate change, including maize, oil crops, pulses, and sugar, no single technology suite can fully counter the multifaceted impacts of climate change at the national scale; and even stacking of technologies will not recoup the productivity levels of a no-climate-change future.

Given that the economic costs to the Egyptian population from climate change impacts on agriculture are estimated at $1.84 billion per year, investments in effective adaptation strategies should be advanced by the government of Egypt, and in particular, investments that address reduced yields and food security impacts.

Egypt's strong linkages with world food markets will allow the country to benefit from adaptation efforts elsewhere in the world, through lower global food prices and increased opportunities for food trade. In the case of a globally accelerated adaptation effort, food prices increase less, reducing pressure on consumers and producers, and food imports are more accessible and can better compensate for domestic shortfalls. In this case, imports of cereals, including wheat, rice, and maize almost return to no-climate-change levels (decreasing by just over 1 percent); for exports, only fruits and vegetables show continued declines.

In sum, Egypt will benefit greatly from accelerating adaptation efforts, particularly if such investments do not increase greenhouse gas emissions. Egypt will also benefit from supporting adaptation (and mitigation) efforts elsewhere in the world. Global advances in adaptation and mitigation make it more likely that Egypt can continue importing food commodities at affordable prices, a strategy that should be pursued given rapid population growth and limited land and water resources.

OTHER INDIRECT COUNTERMEASURES AGAINST CLIMATE CHANGE ALSO NEED FURTHER ANALYSIS

Many other measures, beyond the technologies studied here, can address negative climate change impacts on the agriculture sector. These include development of additional land and water resources (for example, desalination, nonrenewable groundwater); improving the functioning of agricultural markets, including more market-based allocation of resources and removal of input and output subsidies in favor of income transfers; a more climate-impact focused trade response, including a reassessment of which strategic crops to grow in-country and which ones to import; and an economywide assessment of which sectors to strengthen in the face of projected climate change impacts.

INTRODUCTION

The importance of a resilient agriculture sector in providing food security, livelihoods, and household income was highlighted in many countries by the recent pandemic, as was the capacity of the sector to cushion the negative impacts of the subsequent economic slowdown. This has been the case in Egypt, where agriculture has been resilient to the health crisis in comparison with the service and industry sectors (Breisinger et al. 2020). However, the sector's resiliency is gradually being corroded by climate change, with lasting, harmful effects for agriculture and food systems.

The impacts of climate change on the agriculture sector are often insidious, as temperature and precipitation regimes are changing gradually. However, these gradual changes are interspersed with extreme events, such as droughts and flooding, that are increasing in frequency and intensity with climate change. Given that temperatures in Egypt are already high and water resources and fertile lands extremely scarce, climate change poses a significant threat to Egypt's agricultural future.

This report presents 1) an assessment of the impacts of climate change on Egypt's agriculture sector, food security, and nutrition; 2) alternative adaptation options at the farm and national policy levels; and 3) policy recommendations aimed at strengthening the Egyptian food system's resilience to climate change.

1. THE AGRICULTURE SECTOR

LAND AND WATER RESOURCES

Egypt's total land area of 100.15 million ha makes it the 31st largest country in the world but, due to its hot desert climate and extremely limited precipitation, only about 3.89 million ha (3.9 percent) are suitable for agricultural production; as a result, Egypt ranks 97th in agricultural land area. To address aridity, almost all agricultural lands are irrigated; only a small share of agricultural land (< 0.4 percent) along the northern coast is rainfed. A considerable share of agricultural land (25 percent) is planted to permanent crops and 1 percent is planted forest land (Table 1.1).

The areas under different land uses show annual fluctuations, although generally with long-term growth from 1990 to 2020. The exception is forest land, which decreased by 0.14 percent per year. Land planted to permanent crops grew fastest, with an annual growth rate of 3.9 percent, followed by irrigated croplands at 1.1 percent annually. The strip of fertile land around the Nile Valley and Delta (Figure 1.1) is home to 95 percent of the country's population and 85 percent of cultivated areas (the "old lands"), which benefit from Nile River water delivered through a 35,000-km network of irrigation canals.

The Nile is the most important source of water for agriculture, domestic/household use, and industry, providing an annual supply of about 55.5 billion cubic meters (bcm) of water, equivalent to 73 percent of total supply (Table 1.2). Wastewater recycling supplies a total of about 17 percent of water resources, about 15 percent (11.9 bcm) from reuse of agricultural drainage water and 2 percent (1.2 bcm) from treatment of municipal wastewater. Groundwater supply has also increased in recent years, facilitated by cheaper well-drilling technologies, and now provides about 10 percent of the country's irrigation needs by pumping 6.9 bcm of water annually (Table 1.2). Other sources of water are occasional rains and floods and seawater desalination facilities – together contributing about 1 percent of the country's total water supply. Desalinating seawater for irrigation use is not yet being considered, but may be in the future if technology costs decline further.

In terms of water demand, agriculture remains the dominant user, consuming 62.2 bcm of water annually, equivalent to 82 percent of total supply, which exceeds the approximately 55.5 bcm annual water supply from the Nile. Domestic consumption ranks second with 10.4 bcm, followed by industry with 1.2 bcm. With increasing population and investment in industry, these sectors' water needs are expected to continue growing. Water losses to evaporation are also substantial at 2.5 bcm (3 percent).

FIGURE 1.1 SATELLITE IMAGE OF EGYPT: Importance of the Nile River and Delta for agriculture

Source: Google Maps, accessed 2021.

TABLE 1.1 LAND-USE AREA AND CHANGE IN EGYPT, 1990-2020

Land classification	Years				Annual growth rate by period				Change from 1990 to 2020
	1990	2000	2010	2020*	1990-2000	2000-2010	2010-2020	1990-2020	
	'000 hectares				percent				
Country area*	100,145	100,145	100,145	100,145	–	–	–	–	–
Agricultural land	2,648	3,291	3,671	3,886	2.43	1.04	0.65	1.04	47
Arable land	2,284	2,801	2,873	2,929	2.29	-0.49	0.49	0.39	28
Land equipped for irrigation	2,648	3,291	3,664	3,871	2.28	1.02	0.62	1.06	46
Land under permanent crops	364	490	798	958	3.33	7.82	1.26	3.90	163
Forest land (planted)	44	59	66	40	3.05	1.04	-4.69	-0.14	-8

Source: FAO 2020. FAOSTAT dataset.

Note: *Country area includes 600,000 ha of inland waters. Latest data from FAO are for 2019, values for 2020 are trend-extrapolation of preceding 5 years (2015–2019). Although only values for 1990, 2000, 2010, and 2020 are presented in the table, growth rate estimations were done with the use of annual data and a logistic growth function.

TABLE 1.2 SOURCES AND USES OF WATER RESOURCES IN EGYPT, CROP YEARS 2012/13 TO 2015/16

Water Resources	Crop year				Average share
	2012/13	2013/14	2014/15	2015/16	
Source	billion cubic meters				percent
Share of Nile River water	55.5	55.5	55.5	55.5	73
Groundwater in Nile Valley and Delta	6.7	6.7	6.9	6.9	9
Agricultural drainage water recycling	11.2	11.5	11.7	11.9	15
Municipal wastewater recycling	1.3	1.3	1.3	1.2	2
Rain and floods	0.7	0.9	0.9	0.7	1
Seawater desalination	0.1	0.1	0.1	0.1	0
Total	75.5	76.0	76.4	76.3	100
Uses					
Agriculture	62.1	62.4	62.4	62.2	82
Evaporation from the Nile and canals	2.5	2.5	2.5	2.5	3
Domestic uses and water for drinking	9.7	10.0	10.4	10.4	13
Industry	1.2	1.2	1.2	1.2	2
Total	75.5	76.0	76.4	76.3	100

Source: CAPMAS (2019).

Note: Values are net of environmental flows.

AGRICULTURE IN THE EGYPTIAN ECONOMY

Agriculture is a key sector for the Egyptian economy, representing about 11 percent of the country's annual gross domestic product (GDP) and 24 percent of total employment. The service sector, which includes tourism, accounts for the largest share of both GDP and employment, while the industry sector, including construction, accounts for about a third of total GDP and a quarter of employment (Table 1.3).

The share of agriculture in GDP has been declining steadily since the 1990s – from 18.51 percent in 1990 to 11.05 percent in 2019, equivalent to a −0.26 percentage point annual decline for the 30-year period. A similar annual rate of decline, of −0.25 percentage points, was recorded in the past decade (2010–2019). Also declining, from a higher initial rate, is the share of employment in agriculture; it fell from 39.0 percent in 1991 to 23.9 percent in 2019, equivalent to a −0.52 percentage point annual decline. The declines in agriculture's share in GDP and employment have mostly favored the service sector, which has generated increasing shares of GDP and employment over the last three decades; the share of the industry sector has been relatively stable (Table 1.3).

The share of agriculture in GDP and total employment varies considerably across governorates, reflecting differences in agro-climatic conditions and degrees of economic diversification. Lower Egypt has a relatively higher share of GDP stemming from agriculture, while the governorates of Upper Egypt have a higher share of employment in agriculture (Kassim et al. 2018).

Agricultural and food productivity grew rapidly during the past 30 years, with production doubling from 1990 to 2010, and then slower growth from 2015 to 2019. Trajectories of food import and export values reflect the agriculture sector's changing importance relative to other sectors of the economy. While the country's imports of all goods grew at a steep annual rate of 13.8 percent from 2000 to 2010 (compared to food import growth of 10.7 percent), that rate slowed to 4.8 percent per year in the last five years. Growth of food imports did not decline as much, continuing to grow at a modest rate of 5.1 percent. Agricultural exports achieved growth of 26.3 percent per year during the 2000–2010 period, but growth then slowed to just 0.06 percent per year, while exports of all goods achieved 7.2 percent growth per year.

Egypt remains highly vulnerable to international market developments because it relies on imports for key food commodities, including wheat and rice. More than 50 percent of wheat and 15 percent of rice demand are imported even though wheat is the most widely cultivated crop in Egypt, and often grown with rice. These two crops are staple foods and account for significant portions of household food budgets and government cost outlays (that is, direct subsidies and import costs).

A look at Egypt's political economy contextualizes this vulnerability. Broadly speaking, growth of agriculture and food production has been supported by policy reforms that, to varying degrees, have reinforced the role of the state in the economy and affected Egypt's position in the global market. In the 1950s and 1960s, the country transitioned out of feudal and semi-feudal systems, promoting the rural citizen and factory worker to the forefront of post-independence Egyptian identity. It did so by transforming the economy through state-led measures for industrialization and large-scale nationalization, including significant spending on infrastructure, land reform and redistribution, and social services (Alissa 2007). Centralized price controls and controls over marketing, raw materials, agricultural inputs, and foreign currency were put in place. The private sector was also limited to activity in just a few sectors, including agriculture. Foreign trade was largely in the hands of state-owned companies, allowing the government to exercise additional controls on imports. For example, wheat supply was governed by a compulsory delivery policy, accompanied by a reduction in the allocated production area, and sale of a certain quota at prices fixed below international rates (Cassing et al. 2009).

The Open Door Economic Policy period, which began in 1973, saw efforts to open up the Egyptian

TABLE 1.3 AGRICULTURE AND THE EGYPTIAN ECONOMY, 1990, 2000, 2010, 2015-2019

		Decades			Years				
Indicators	Units	1990	2000	2010	2015	2016	2017	2018	2019
GDP									
Gross domestic product (GDP)	constant 2010 US$ billion	87	136	219	250	261	272	286	302
GDP growth	annual rate (%)	5.67	6.37	5.15	4.37	4.35	4.18	5.31	5.56
Agriculture, forestry, and fishing	% of GDP	18.51	15.54	13.34	11.39	11.77	11.49	11.23	11.05
Agriculture, forestry, and fishing	annual rate (%)	2.92	3.70	3.47	3.07	3.10	3.24	3.11	3.35
Industry (including construction)	% of GDP	27.41	30.75	35.78	36.63	32.46	33.75	34.96	35.62
Services	% of GDP	49.63	46.53	46.23	53.17	54.48	53.24	51.50	50.47
Food Production and Trade									
Agriculture production	index: 2014–2016 = 1.00	0.45	0.71	0.90	1.01	0.98	1.01	0.97	1.02
Food production	index: 2014–2016 = 1.00	0.43	0.70	0.90	1.01	0.99	1.01	0.97	1.03
Imports									
Imports of goods	current US$ million	9,216	14,578	52,923	63,574	55,789	61,627	72,000	70,919
Food imports	current US$ million	–	3,674	10,129	12,359	11,370	12,504	13,198	14,694
Exports									
Exports of goods	current US$ million	2,585	5,276	26,438	21,349	25,468	25,604	27,624	28,993
Food exports	current US$ million	–	422	4,373	4,402	5,157	4,509	4,362	4,801
Employment									
Employment in agriculture	% of total employment	39.0*	29.6	28.3	25.8	25.6	25.0	24.3	23.8
Employment in industry	% of total employment	21.3*	21.3	25.4	25.1	25.5	26.6	27.2	27.7
Employment in services	% of total employment	39.6*	49.1	46.3	49.1	48.9	48.4	51.5	52.4
Population and Income									
Total population	million	56.1	68.8	82.8	92.4	94.4	96.4	98.4	100.4
Population growth	annual rate (%)	2.43	1.93	1.98	2.21	2.15	2.09	2.03	1.98
GDP per capita	constant 2010 US$	1,558	1,982	2,646	2,705	2,763	2,819	2,909	3,010
GDP per capita growth	annual rate (%)	3.13	4.34	3.08	2.09	2.13	2.03	3.19	3.49

Source: Data from World Development Indicators.

Note: Production indices are calculated from the underlying volume of production or per capita production, normalized to the base period 2014–2016. Employment values are modeled ILO estimates – means data not available; * means values for 1990 not available, values shown are for 1991 instead. For brevity, some calculations and estimates based on this table (e.g., growth rates and percentages) are presented and discussed in the text without inclusion in the table.

economy to Arab and other foreign investment, reduce control over private sector activity, and establish "free zones" offering foreign private firms tax holidays and import and export licenses. While some liberalization of the economy was achieved, large increases in imports began to hurt domestic production, including food products, such as wheat (Alissa 2007). Attempts to raise the price of bread by removing subsidies were met with riots (Ibrahim and Ibrahim 2003). The compulsory delivery quotas for wheat were then replaced with optional delivery requirements, but still at fixed prices (Kassim et al. 2018). Nevertheless, demand for wheat continued to exceed supply. Since the 1980s, Egypt has consistently been one of the world's top importers of wheat.

The period after 1985, coinciding with the global oil price crash, began with another expansion in public sector expenditures, primarily through subsidies. However, subsequent reform periods sought to reduce the role of the state in the economy through liberalization, privatization, and market-based principles, essentially opening the Egyptian economy up to global competition while encouraging exports, and enhancing productivity and domestic income growth. This included the Agricultural Production and Credit Project (1987–1995) and the Agricultural Policy Reform Program (1996–2002), both of which reduced subsidies on agricultural inputs, liberalized markets, removed controls on prices and marketing restrictions for major crops, privatized public firms, and encouraged further private sector participation in agricultural production and trade (Baffes and Gautam 1996; Cassing et al. 2009; Gouda 2016). During this period, the government's guaranteed floor prices were fixed close to international prices, thus setting distortionary price supports at the minimum.

A focus on trade and institutional measures also marked this period. Egypt reduced the maximum tariffs on products to 40 percent in 1998. Nontariff barriers were also addressed, and privatization was accelerated, continuing into the late 2000s. Moreover, Egypt signed a number of trade agreements, including with the United States, European Union, Eastern and Southern Africa, and other Arab countries. These agreements pushed a more sensitive approach to meeting international standards, especially in the agriculture and industry sectors (Alissa 2007), including standards related to production, phytosanitary measures, quality, compliance, technology use, export processes, packaging, data recording, and maritime and other transport (Sultan 2020).

More recently, in the past decade, the policy of shifting production to high-value and less water-intensive export crops and of importing water-intensive and less-expensive food crops, was further applied to the rice sector. Moreover, Egypt has, for a long time, restricted water use for rice production by limiting the area for rice cultivation. More recently, rice exports have been restricted and the rice import tariff was reduced to 5 percent (Sultan 2020). For wheat, however, the government continued to be the largest buyer of local wheat production, offering farmers procurement prices set above world prices to encourage domestic production (Kassim et al. 2018). But dependence on imported food remains high, especially for wheat.

This continued state engagement can be explained by the strategic importance of the agriculture sector in Egypt. Ministerial Decree 215 for 2014 considered the impact of food price volatility on the income of the poor as a major argument for continuing consumer food subsidies through the Tamween ration card program. Presidential Decree 25 of 2016, in tandem with economic reforms that included the devaluation of the Egyptian currency, sought to limit outflow of foreign currency by once again raising tariff rates on a variety of imports, including agricultural and food items. Tariffs on apples, grapes, and pears were raised by as much as 40 percent and tariffs on nuts by as much as 20 percent (Sultan 2020). In addition, the agriculture sector's disproportionate role in water use and water security reinforces the tight political grip of the government on the sector.

High food production growth rates over the past three decades are reflected in high annual growth rates for livestock products, some food crops, and food exports. Among livestock products, production of poultry rose 5.8 percent annually from 1990 to 2020, eggs 4.4 percent, and milk 3.3 percent (Table 1.4). In the food crop sector, roots and tubers (including potatoes) have grown fastest, with annual growth of 4.3 percent, followed by fruits at 3.5 percent and vegetables at 2.9 percent. Food exports accounted for 20 percent of total exported goods over the last five years, including 3.4 million metric tons (mt) of fruits and vegetables in 2020 (12 times the volume in 2000), while the country imports lower-value cereals and pulses, accounting for some 21 percent of total imports. This strategy saves scarce water and land resources (Table 1.1 and Table 1.4).

Among cereals, wheat achieved the highest production growth rate at 2.9 percent, followed by maize at 1.8 percent annually since 1990 (Table 1.4). However, growth rates have been slowing or

TABLE 1.4 RECENT PERFORMANCE OF EGYPT'S AGRICULTURE SECTOR – FOOD PRODUCTION GROWTH, 1990-2020

Commodity/Indicators	Years				Annual growth rate by period				Change from 1990 to 2020
	1990	2000	2010	2020	1990-2000	2000-2010	2010-2020	1990-2020	
Production		'000 mt					percent		
All Meat Products	721	1,331	1,831	2,440	6.21	3.09	2.90	3.80	239
Beef/Buffalo meat	298	557	829	747	5.69	4.16	-1.14	2.86	150
Sheep/Goat meat	84	122	140	76	4.65	1.66	-5.21	0.49	-10
Poultry meat	258	555	791	1,550	8.43	3.17	6.91	5.77	500
Eggs	157	166	299	392	1.76	5.28	2.28	4.39	151
Milk	2,233	3,663	5,779	5,012	5.33	5.21	-1.56	3.29	124
All Cereals	11,306	19,157	21,988	20,971	4.87	1.87	-0.28	1.94	85
Maize	4,472	6,318	7,376	6,799	2.87	1.42	-0.44	1.76	52
Rice, milled	1,774	3,622	3,803	3,184	6.95	1.51	-1.78	1.23	79
Wheat	3,429	6,335	7,892	8,405	5.78	2.83	0.79	2.87	145
Fiber crops*	20	16	11	10	-1.93	-2.28	-1.43	-1.97	-51
Fruits	5,659	9,004	12,244	15,265	5.03	3.37	2.55	3.54	170
Nuts	5	13	43	25	5.83	8.82	-5.22	8.21	395
Vegetables	7,624	11,609	17,617	15,284	4.55	4.60	-1.33	2.91	100
Oil crops	1,088	1,151	1,087	1,935	1.10	-0.34	5.28	0.95	78
Pulses	510	464	360	287	-0.51	-3.00	-1.72	-2.59	-44
Roots and tubers**	1,917	2,156	4,085	5,617	2.59	7.12	2.74	4.28	193
Sugar crops	11,696	17,571	21,989	26,245	3.99	2.26	1.90	2.97	124

Source: Data from FAOSTAT, accessed 2020.

Note: Values are 3-year moving averages of t-2 t-1 and t0. Latest data from FAO are for 2019 only, values for 2020 are trend-extrapolation of preceding 5 years (2015-2019), before the 3-year moving averages were applied to 2020; *Though not food per se, seeds of fiber crops are sources of oils and meals. **Mainly potatoes.

declining sharply over the last decade, with roots and tubers, fruits, eggs, sugar crops (cane and beets), and wheat displaying slowing growth, and production of sheep and goat meat, nuts, rice, milk, vegetables, and beef production contracting. Notable exceptions are poultry and oil crops, which have posted increasing rates in the last 10 years, and fiber crops and pulses, with negative growth rates for the entire 1990-2020 period (Table 1.4, Figure 1.2).

Patterns of change in harvested areas and yields are similar to the production growth patterns, but with faster declines for harvested areas (Appendix Tables A1 and A2). Trends in production as well as area and harvested yields are depicted in Figures 1.2 and 1.3. Declining production performance, coupled with increasing demand from population pressure, has led to increased imports for major commodity groups (Table 1.5), making Egypt one of the largest wheat importers in the world and a large net importer of maize, pulses, and oil crops in the most recent decade.

FIGURE 1.2 TRENDS IN PRODUCTION OF MAJOR FOOD COMMODITIES IN EGYPT, 1990-2020

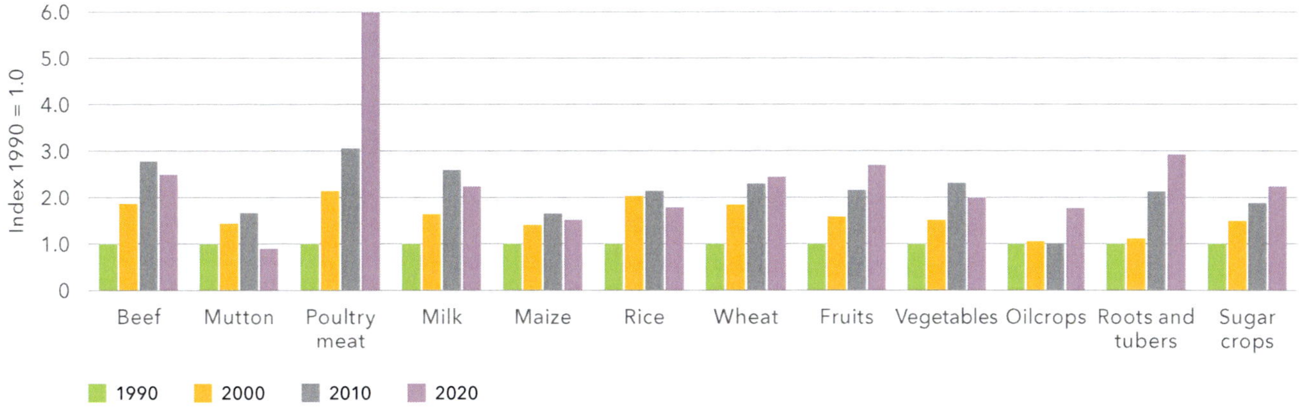

Source: FAOSTAT, accessed 2020.

Note: Values are 3-year moving averages of t-2 t-1 and t0. Since latest data from FAO are for 2019 only, values for 2020 are trend-extrapolation of 5 preceding years (2015–2019), before the 3-year moving averages were applied.

TABLE 1.5 TRADE PERFORMANCE OF EGYPTIAN AGRICULTURE, 1990-2020

Commodity/ Indicators	Years			
	1990	2000	2010	2020
Net Trade		*'000 mt*		
Meats				
Beef/Buffalo meat	-130	-172	-144	-747
Mutton/Goat meat	-5	-1	-4	-1
Poultry meat	-14	-2	-67	-33
Eggs	-3	-1	-2	1
Dairy products	-119	-102	-56	-71
Cereals				
Maize	-1,543	-4,130	-5,179	-7,330
Rice, milled	51	374	493	-536
Wheat	-5,444	-4,856	-9,346	-10,007
Textile fibers	9	72	9	-148
Fruits	154	89	808	2,480
Nuts	0	-8	6	29
Vegetables	88	179	421	932
Oil crops	-46	-224	-1,472	-4,508
Pulses	-50	-232	-354	-5,120
Roots and tubers	98	163	256	529
Sugar/Other sweeteners	-483	-762	-631	-160

Source: Data from FAOSTAT, accessed 2020.

Note: Values are 3-year moving averages of t-2 t-1 and t0. Since latest data from FAO are for 2019 only, values for 2020 are trend-extrapolation of 5 preceding years of 2015–2019, before the 3-year moving averages were applied. Negative values mean net imports; positive values mean net exports.

FIGURE 1.3 PRODUCTION, AREA HARVESTED, AND YIELD PERFORMANCE OF SELECTED FOOD CROPS IN EGYPT, 1990–2020

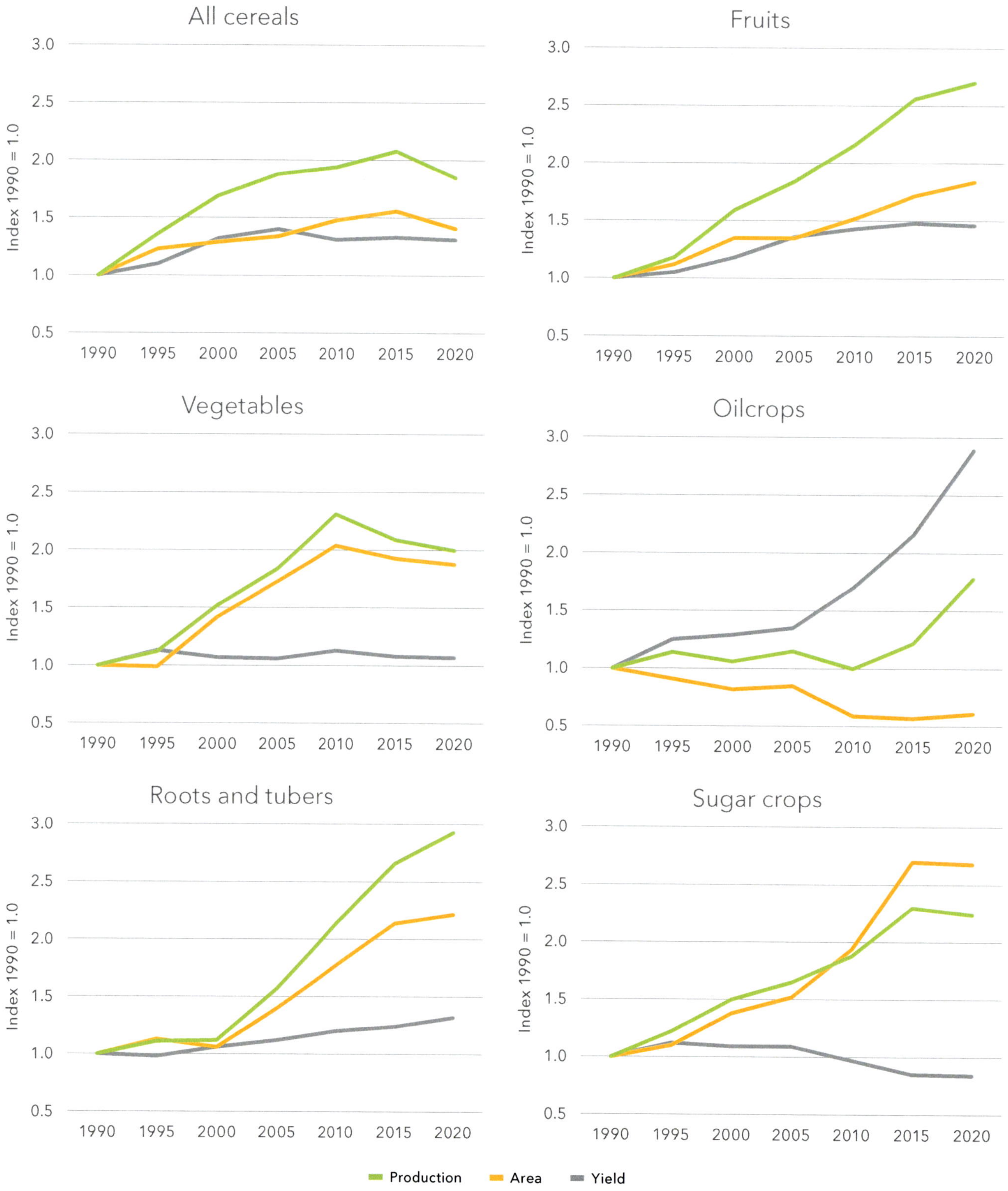

Source: Data from FAOSTAT.

2. CLIMATE CHANGE AND RESILIENCY OF THE AGRICULTURE SECTOR

Egypt's agriculture sector is particularly affected by climate and climate change. Climate change affects crop productivity through changes in rainfall and temperature patterns, both of which directly influence the supply of water, the demand of crops for water, as well as heat stress experienced by crops. Livestock and fish are likewise directly affected by heat stress and variability of water supply, and indirectly through the effects of climate change on the production of feeds.

BIOPHYSICAL-ECONOMIC MODELING

The International Model for Policy Analysis of Agricultural Commodities and Trade (IMPACT),[1] depicted in Figure 2.1, is the main modeling framework used in this study to estimate the impacts of climate change on the agriculture sector and to determine the effectiveness of adaptation policies designed to counter them. IMPACT combines biophysical models (climate, hydrology, and crop growth) with economic models to project water and food supply and demand as well as food trade and prices under climate change. The water models, informed by the climate models, estimate the changes in the supply of water from various sources and allocate available supply to different users, including households, industry, livestock, irrigation, and the environment. The IMPACT economic model simulates national and global markets of agricultural production, demand, and trade that are associated with 62 agricultural commodities across 158 countries and regions. The market-clearing

FIGURE 2.1 BIOPHYSICAL AND ECONOMIC MODELING FRAMEWORK OF THE IMPACT MODEL

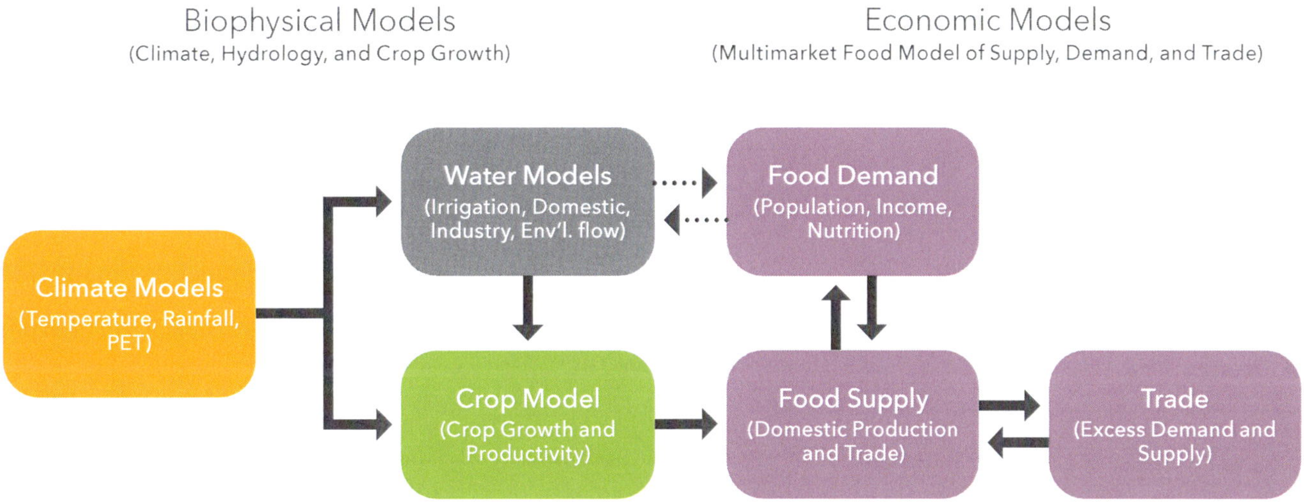

Source: Constructed by the authors.

1 For more about IMPACT, see https://www.ifpri.org/project/ifpri-impact-model.

iterative equilibration of supply and demand, at the country and global levels, determine world prices of the agricultural and food commodities.

ALTERNATIVE CLIMATE FUTURES

Climate models

Historical records confirm the gradual and sustained increase in temperatures in Egypt. Over a 65-year period, the average winter temperature increased by 0.5–0.85°C, while summer temperatures increased by 0.45–0.7°C across Egypt (Omran 2020).

Three general circulation models (GCMs) of climate from CMIP5[2] were used to simulate future climate conditions. These are GFDL from the Geophysical Fluid Dynamics Laboratory of Princeton University in the United States; HGEM, the Global Environmental Model of Met Office Hadley Centre in the United Kingdom; and the IPSL model of the Institut Pierre-Simon Laplace in France. All simulations were done under the representative concentration pathways (RCP)/shared socioeconomic pathways (SSP) framework, applying RCP 8.5 emissions of greenhouse gases (GHG) and SSP2 population and GDP assumptions (IIASA 2013, 2015). RCP 8.5 represents the highest level of emissions among the RCPs, but is also reflective of overall growth in actual emissions.

Temperatures are predicted to continue increasing and to influence the patterns of rainfall and potential evapotranspiration (PET). Changes in mean daily maximum temperature for the Nile Basin countries, as projected by the 3 GCMs, are shown in Figure 2.2. Annual temperature increases for Egypt are as large as 3.1°C for HGEM and projected to be highest for the July-August-September period (Table 2.1). Annual rainfall is projected to decline, with the largest reduction of 15 mm for IPSL, while PET is expected to increase by 5–21 percent.

Economic models

The biophysical effects of climate change are transmitted to IMPACT's economic module, which simulates as outputs global and national production, area,

TABLE 2.1 SEASONAL AND ANNUAL CLIMATE FOR EGYPTIAN AGRICULTURAL AREAS, 1970–2000, AND CHANGES THROUGH 2040–2070 FOR 3 CLIMATE SCENARIOS FROM CMIP5

Season/ Months	Baseline, 1970–2000	Change from baseline, 1970–2000 to 2040–2070		
		GFDL	HGEM	IPSL
	Mean daily maximum temperature, °C			
Jan-Feb-Mar	21.2	1.7	2.8	3.2
Apr-May-Jun	31.2	1.9	3.2	2.8
Jul-Aug-Sep	33.7	2.2	3.7	2.9
Oct-Nov-Dec	25.4	2.5	2.8	3.2
Annual	27.9	2.1	3.1	3.0
	Mean daily minimum temperature, °C			
Jan-Feb-Mar	7.8	1.8	2.7	2.6
Apr-May-Jun	16.3	1.8	3.3	2.3
Jul-Aug-Sep	20.5	2.3	4.1	2.9
Oct-Nov-Dec	12.9	2.5	3.4	2.6
Annual	14.4	2.1	3.4	2.6
	Rainfall, millimeters			
Jan-Feb-Mar	21	0	-7	-8
Apr-May-Jun	2	0	-1	1
Jul-Aug-Sep	0	0	0	0
Oct-Nov-Dec	19	-5	-4	-8
Annual	42	-6	-12	-15
	Potential evapotranspiration, millimeters			
Jan-Feb-Mar	288	14	26	38
Apr-May-Jun	604	31	45	52
Jul-Aug-Sep	572	25	36	38
Oct-Nov-Dec	286	19	11	32
Annual	1750	89	117	160

Source: Based on Müller and Robertson (2014).

Note: GFDL = Geophysical Fluid Dynamics Laboratory; HGEM = Hadley Centre Global Environmental Model; IPSL = Institut Pierre-Simon Laplace. Simulations are based on Representative Concentration Pathway 8.5.

2 The Intergovernmental Panel for Climate Change (IPCC) regularly publishes comprehensive scientific assessment reports (ARs) on climate change, its causes, potential impacts, and response options. It also includes updated climate models and estimates of alternative climate futures. In the IPCC-AR5, several climate models form part of the Coupled Model Intercomparison Project Phase 5 (CMIP5), from which the 3 GCMs used in this study were selected.

FIGURE 2.2 PROJECTED CHANGE IN MEAN DAILY MAXIMUM TEMPERATURE (°C), 1970-2000 TO 2040-2070

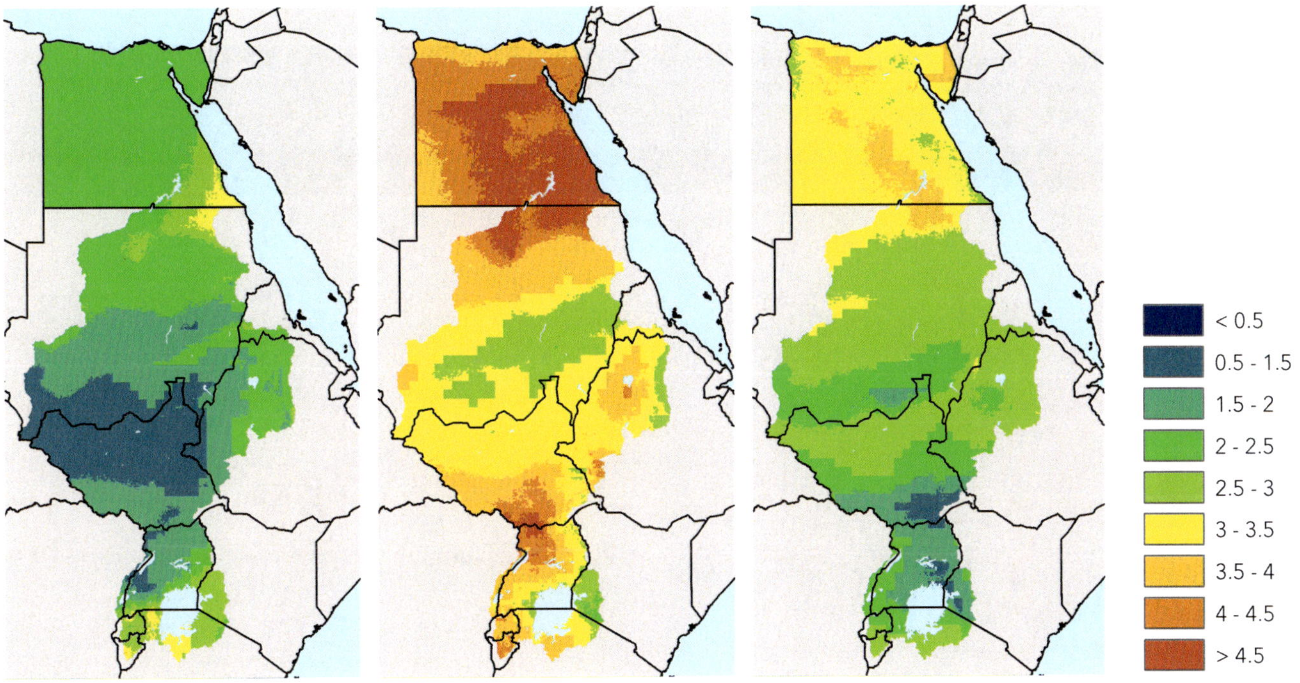

Source: Based on Müller and Robertson (2014).

Note: Left to right, GFDL, HGEM, and IPSL. GFDL = Geophysical Fluid Dynamics Laboratory; HGEM = Hadley Centre Global Environmental Model; IPSL = Institut Pierre-Simon Laplace. Simulations are based on Representative Concentration Pathway 8.5.

yield, trade, demand, and prices for agricultural commodities (Robinson et al. 2015). Because of the model's capacity to simulate the impacts and estimate the costs of climate change, it has been used extensively in policy analysis for climate change adaptation and mitigation; in studies of the impacts of agricultural technology developments and irrigation investments; and for projections of food supply and demand, trade, and food security to 2050.

The final IMPACT projections are presented as the mean of outputs derived from running individual simulations for each of the three climate scenarios.

CHANGING TRAJECTORIES OF AGRICULTURAL PRODUCTIVITY AND FOOD SECURITY

Changes in temperature and rainfall patterns brought on by climate change alter crop yields both directly and indirectly via changes in water availability for irrigation. Livestock productivity is indirectly affected through changes in feed availability. Direct heat stress on livestock is not considered in the model.

The biophysical and economic effects of climate change are presented in Table 2.2. The biophysical effects were estimated with the GFDL, HGEM, and IPSL climate models (temperature, rainfall, and PET), the DSSAT crop model (temperature stress), and the IMPACT water module (availability of irrigation water and water stress), while estimates of stress from increased salinity are taken from the literature (Hammam and Mohamed 2020; Kotb et al. 2000). The IMPACT economic model simulates the impacts of producer and consumer responses to climate-induced changes in market prices, which affect food supply and demand decisions.

Yield growth in a no-climate-change scenario is projected to slow in the 2020–2050 period

TABLE 2.2 CHANGES IN PRODUCTIVITY DUE TO BIOPHYSICAL AND ECONOMIC EFFECTS OF CLIMATE CHANGE, EGYPT, PROJECTED BY 2050

	Biophysical effects				Combined biophysical and economic effects	
Commodities	Heat stress	Water stress	Salinity	Cumulative effects	Egypt	Rest of world
Productivity	*% change from no-climate-change scenario*					
All food crops	-4.94	-4.14	-1.55	-10.29	-6.17	-5.24
All cereals	-4.66	-2.57	-1.59	-8.59	-10.36	-7.74
Maize	-12.86	-2.46	-1.36	-16.16	-19.54	-17.66
Rice	-5.81	-1.59	-1.58	-8.78	-8.53	-5.61
Wheat	2.27	-3.25	-1.78	-2.81	-0.56	0.82
Fruits and vegetables	-4.73	-5.88	-1.48	-11.66	-8.28	-1.95
Oil crops	-6.98	-3.18	-1.53	-11.31	-12.08	-6.69
Pulses	-5.46	0.04	-1.57	-6.92	-9.98	0.01
Roots and tubers	2.61	-0.29	-1.79	0.47	3.56	-4.58
Sugar crops	-6.66	-4.19	-1.56	-11.96	-13.28	-10.39

Source: Estimates derived from IMPACT results.

Note: Values are averages of 3 GCMs: GFDL = General Fluid Dynamics Laboratory; HGEM = Hadley Centre Global Environmental Model; IPSL = Institut Pierre-Simon Laplace; all RCP 8.5 and SSP2.

compared to earlier years, but without any major drop-offs. Yield growth is expected to be relatively stable for maize and wheat, and slow faster for barley, millet, and sorghum (Appendix Table A3).

Under climate change, yields for food crops are projected to decline by 10 percent by 2050, as a result of heat stress (4.9 percent), water stress (4.1 percent), and salinity (1.6 percent). The highest biophysical yield declines are estimated for maize (−16.2 percent), sugar crops (−12.0 percent), and fruits and vegetables (−11.7 percent). Among the biophysical yield stressors, temperature contributes the most, at −12.9 percent for maize, −7.0 percent for oil crops, and −6.7 percent for sugar crops. The yield decline is less for wheat (−2.8 percent), while biophysical impacts result in a small yield increase for roots and tubers.

The combined biophysical and economic impacts of climate change on crop yields are projected to be more severe for Egypt than the rest of the world (ROW). The exception is roots and tubers, where an overall positive yield increase is achieved, at 3.6 percent (Table 2.2 and Figure 2.3). The positive impact for potatoes is due to positive biophysical impacts compared to ROW, that is, improved yields from higher temperatures, as well as the endogenous productivity effect from higher global food prices resulting from climate change.

The largely negative impacts of climate change on crop yields translate into production declines, which are again larger for Egypt (−5.7 percent) compared to ROW (−4.4 percent) (Table 2.3). The largest production declines are projected for maize, both for Egypt (−21.8 percent) and ROW (−22.1 percent), while the spread is widest for pulses, at −23.9 percent for Egypt and only −0.23 percent for ROW.

Estimates of the impacts of climate change on the production of animal-source foods (meat, dairy, and eggs) were derived indirectly from its impacts on animal feed and are thus conservative. The model does not include any estimate of the climate-induced heat stress on animals that may negatively affect productivity.

Declines in yields and production are projected to push world prices of food up by as much as

FIGURE 2.3 CHANGES IN YIELD TRAJECTORIES DUE TO COMBINED BIOPHYSICAL AND ECONOMIC EFFECTS OF CLIMATE CHANGE FOR MAIZE, RICE, AND WHEAT, PROJECTED, 2020–2050.

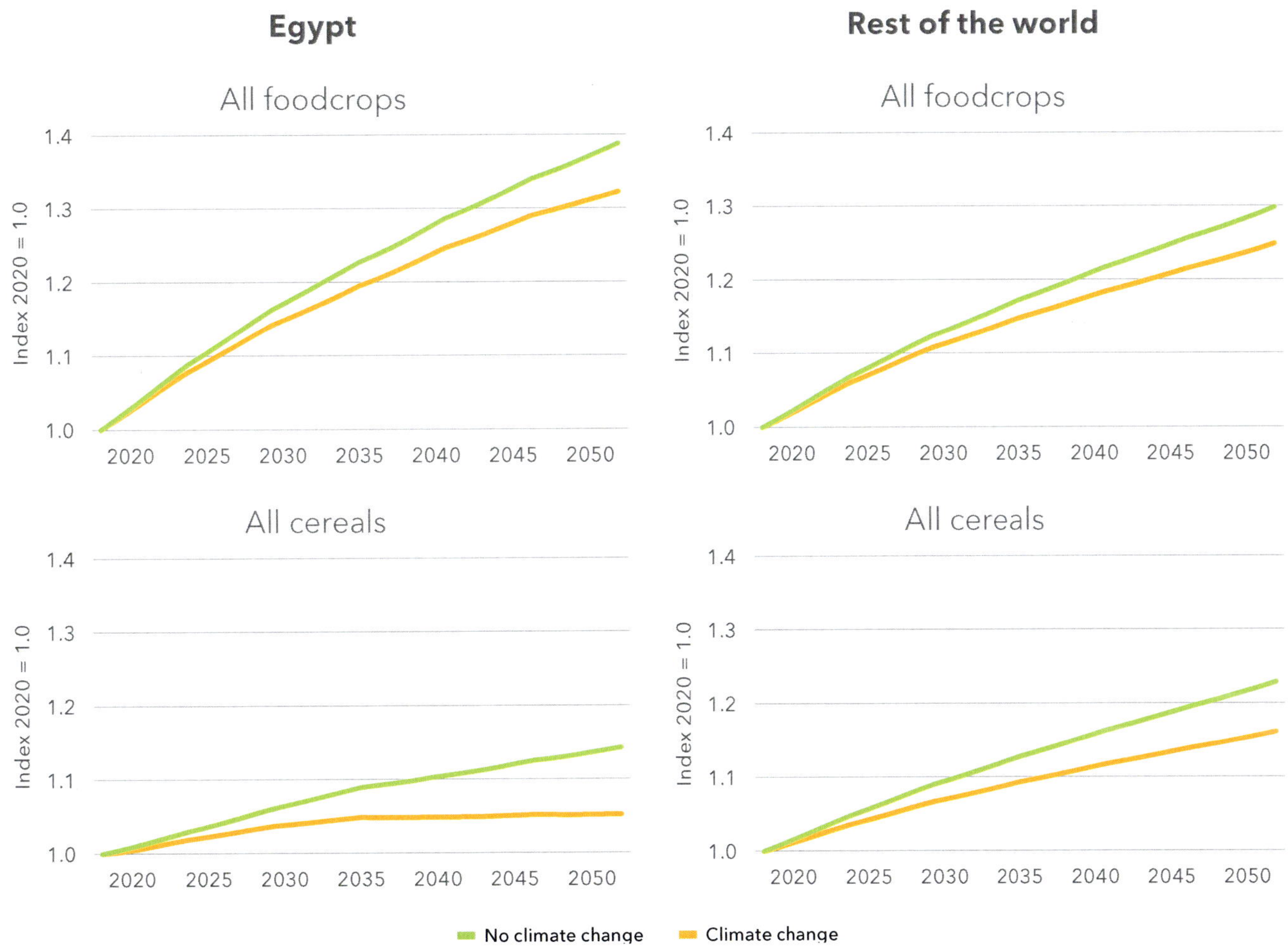

Source: IMPACT simulations.

Note: Values for climate change scenarios are averages of projections from GFDL, HGEM, and IPSL climates, normalized with 2020 values (2020 = 1.00). Annual averages and ranges of yields for the 3 climate scenarios were both estimated. For 2050, under climate change, for all food crops for Egypt, normalized yield averages: 1.3228 with a range from 1.2892 to 1.3484; for ROW normalized yield averages 1.2486 with a range from 1.2341 to 1.2600. For all cereals for Egypt, normalized yield averages 1.0523 with a range from 1.0448 to 1.0640; for ROW yield averages 1.1861 with a range from 1.1346 to 1.2286.

22.8 percent for maize and by 18.7 percent for rice (Figure 2.4). Prices of oil crops, roots and tubers, and poultry are also projected to increase significantly. Higher food prices, induced by climate change, will pose a significant threat to food security and limit access to food for the poorer segments of society. Higher food prices will also reduce Egypt's food import demand by 1.329 million mt (4.5 percent) and dampen the demand for Egypt's exports, for example, of fruits and vegetables (Table 2.4). For Egypt, the global declines in yields and food production will mean tighter food markets that make it more difficult for Egypt to rely on food imports to augment domestic supplies.

The threat of climate change to food security is evident in the projected declines in per capita food and calorie availability. Per capita consumption of food crops is projected to decline by 2.1 percent by

TABLE 2.3 PROJECTED IMPACT OF CLIMATE CHANGE ON FOOD PRODUCTION IN EGYPT AND THE REST OF THE WORLD, 2050

	Egypt			Rest of the world		
Food commodities	No climate change	With climate change	Change	No climate change	With climate change	Change
Production	*'000 mt*		*percent*	*million mt*		*percent*
All meat products	5,721	5,741	0.4	437	433	-0.8
Beef	2,686	2,679	-0.3	118	118	-0.1
Lamb	277	275	-0.4	33	33	0.1
Poultry	2,754	2,783	1.1	158	155	-1.5
Dairy	11,855	11,765	-0.8	1,105	1,103	-0.2
Eggs	660	639	-3.1	90	89	-0.9
All food crops	156,594	147,629	-5.7	13,111	12,538	-4.4
All cereals	27,238	24,003	-11.9	3,239	2,945	-9.1
Maize	12,407	9,697	-21.8	1,293	1,008	-22.1
Rice	2,764	2,586	-6.4	547	527	-3.8
Wheat	10,909	10,661	-2.3	890	890	-0.1
Fruits and vegetables	57,845	52,436	-9.4	2,972	2,913	-2.0
Oilseed crops	1,302	1,228	-5.7	1,349	1,335	-1.0
Pulses	681	518	-23.9	124	123	-0.2
Roots and tubers	11,424	11,774	3.1	1,189	1,138	-4.2
Sugar	4,913	4,739	-3.5	315	309	-1.9

Source: IMPACT simulations.

Note: Only the indirect impacts of climate change on livestock production were estimated through its effects on livestock feeds. Values are averages of 3 GCMs: GFDL = General Fluid Dynamics Laboratory; HGEM = Hadley Centre Global Environmental Model; IPSL = Institut Pierre-Simon Laplace; all RCP 8.5 and SSP2.

2050, driven by reduced consumption of roots and tubers (−8 percent), oil crops (−7 percent), and rice (−5 percent). These are equivalent to a 1.7 percent decline in daily calorie availability (Table 2.5).

The projected reduction in food availability for consumption can worsen children's nutrition outcomes and increase the number of people at risk of hunger. A study for developing countries determined that a 1 percent decline in daily calorie availability can result in a 5.2 percent increase in the number of people at risk of hunger and a 1.3 percent increase in the number of undernourished children (Perez and Rosegrant 2018). Based on Table 2.5, these values would translate into a 9.0 percent increase in the number of people at risk of hunger, and a 2.3 percent increase in the number of undernourished children.

ECONOMIC COSTS OF CLIMATE CHANGE TO AGRICULTURE

Climate change affects both agricultural production levels and food security, through world prices, net trade, and consumption. To get a better estimate of the overall impact, it is helpful to translate all the impacts, negative and positive, into financial terms, since any adaptation to climate change requires monetary outlay, whether from government, producers, or consumers. Understanding these effects of climate change can enable governments and society in general to weigh the costs of climate change adaptation against the projected damages from climate change.

The economic surplus framework for cost-benefit analysis (CBA) as described by Alston, Norton, and

FIGURE 2.4 PROJECTED EFFECTS OF CLIMATE CHANGE ON GLOBAL FOOD PRICES, 2050

Source: IMPACT simulations (see also Appendix Table A4).

Note: No CC = no-climate-change scenario.

TABLE 2.4 IMPACT OF CLIMATE CHANGE ON FOOD TRADE, EGYPT, PROJECTED BY 2050

Food commodities	Net trade of food		
	No climate change	With climate change	Change*
Annual net trade		'000 mt	
All meat products	661	720	59
Beef/Buffalo meat	-31	-34	-4
Mutton/Goat meat	54	54	-1
Poultry meat	637	700	63
Dairy products	3,068	3,017	-50
Eggs	2	-13	-15
All food crops	-29,139	-27,810	1,329
All cereals	-35,957	-32,770	3,187
Maize	-17,609	-14,231	3,378
Rice, milled	-2,868	-2,770	98
Wheat	-12,373	-12,558	-186
Fruits and vegetables	12,387	7,777	-4,610
Oilseed crops	-1,105	-1,025	80
Pulses	-1,653	-1,843	-190
Roots and tubers	4,263	7,244	2,981
Sugar/Other sweeteners	-2,077	-2,076	1

Source: IMPACT simulations.

Note: * Negative values mean declines in exports or increases in imports, while positive values mean increases in exports or declines in imports. Values are averages of 3 GCMs: GFDL = General Fluid Dynamics Laboratory; HGEM = Hadley Centre Global Environmental Model; IPSL = Institut Pierre-Simon Laplace; all RCP 8.5 and SSP2.

TABLE 2.5 CHANGES IN PER CAPITA FOOD AND DAILY CALORIE AVAILABILITY, EGYPT, BY 2050

Food commodities	Consumption		
	No climate change	With climate change	Change
Annual consumption	*kg/capita/year*		*percent*
All meat products	40.5	40.2	-0.75
Beef/Buffalo meat	21.8	21.7	-0.15
Mutton/Goat meat	1.8	1.8	-0.32
Poultry meat	17.0	16.7	-1.57
Dairy products	45.9	45.9	-0.15
Eggs	4.1	4.0	-0.80
All food crops	684.5	670.2	-2.09
All cereals	235.2	231.7	-1.52
Maize	55.4	54.1	-2.32
Rice, milled	35.6	33.7	-5.38
Wheat	140.9	140.6	-0.23
Fruits and vegetables	324.4	318.7	-1.74
Oilseed crops	9.1	8.5	-6.57
Pulses	13.9	14.1	1.57
Roots and tubers	45.3	41.8	-7.66
Sugar/Other sweeteners	49.4	48.1	-2.48
	kcal/capita/day		*percent*
Daily calorie consumption	3,789	3,723	-1.74

Source: IMPACT simulations.

Note: Values are averages of 3 GCMs: GFDL = General Fluid Dynamics Laboratory; HGEM = Hadley Centre Global Environmental Model; IPSL = Institut Pierre-Simon Laplace; all RCP 8.5 and SSP2.

Pardey (1995), commonly used for long-term government infrastructure projects, was used to estimate the long-term economic impact of climate change on the welfare of agricultural producers and consumers, and hence on society as a whole. It measures the costs and benefits that accrue to society from projected shifts in supply and demand over time due to climate change.

Within the economic surplus framework, positive values indicate net benefits while negative values indicate net costs. On this basis, the economic cost of climate change based on changes in the supply and demand of agricultural commodities worldwide is calculated as US$2.43 trillion at the global level for the 30-year period from 2020 to 2050 based on IMPACT simulation results (Table 2.6).[3] Consumers bear the brunt of the costs by paying higher prices for food and other agricultural commodities, incurring overall welfare losses of $4.89 trillion. Producers, on the other hand, register a net gain of $2.47 trillion overall, because the higher prices for their produce, generated by the global effects of climate change, are projected to offset declines in production.

The economic costs to the Egyptian population are estimated at $55.28 billion for the 2020-2050 period, or $1.84 billion per year. Egyptian consumers bear most of the costs of climate change, which amount to $66.25 billion for the 30-year period or $2.21 billion per year, while producers gain an average of $1.97 billion overall, or $0.37 billion per year, implying that higher prices more than compensate for productivity losses. However, most farmers – and especially smallholder farmers – are marginal producers and net buyers of food, and thus are expected overall to suffer economic losses from the combined producer and consumer effects. Moreover, farmers will likely incur higher costs in trying to maintain competitiveness as productivity levels fall, for example, through increased investments in advanced irrigation technology or increased purchase of pesticides to deal with greater pest and disease incidence.

TABLE 2.6 CHANGES IN SOCIETY'S WELFARE DUE TO IMPACTS OF CLIMATE CHANGE ON AGRICULTURE, WORLD AND EGYPT, 2020-2050.

Climate models	Welfare measure		
	Producer surplus	Consumer surplus	Economic surplus
	net present value (US$ billions)*		
World	**2,468**	**-4,894**	**-2,427**
Egypt	**10.97**	**-66.25**	**-55.28**
Annual value	0.37	-2.21	-1.84

Source: IMPACT simulations.

Note: *3% real discount rate applied. Values are averages of 3 GCMs: GFDL = General Fluid Dynamics Laboratory; HGEM = Hadley Centre Global Environmental Model; IPSL = Institut Pierre-Simon Laplace; all RCP 8.5 and SSP2.

3 Throughout this report, "$" refers to US dollars.

3. DEVELOPING RESILIENCE AGAINST CLIMATE CHANGE

In addition to providing researchers and policymakers with comprehensive information on the potential impacts of climate change, the IPCC-AR5 (IPCC 2014) also provided response options to adapt to and mitigate its effects. The various response options can be classified as follows: 1) technological responses, such as developing new crop varieties adapted to changing climate conditions and crop protection technologies; 2) agronomic adaptation responses, such as agroforestry and improved soil and water management practices; 3) improvements in the enabling environment for smallholder agriculture, such as enhancing accessibility to credit facilities, diversification of livelihood practices, and providing early warning systems; and 4) institutional responses, such as strengthening services at local, regional, and national levels, introducing gender-responsive policies, changing trade policies and stockholding policies, and developing other institutions that support adaptation and mitigation.

This study focuses on technological responses to counter climate change, including the development, adoption, and scaling-up of technologies that are resilient to climate change, increase farm profitability, promote food security, and improve society's welfare. Technologies are expected to originate from both the public and private research systems, with public R&D focused on key cereal crops. We will argue in this report that the government of Egypt should consider expanding public agricultural research investments to a larger set of crops, as some of the country's niche crops are either particularly affected by climate change or will have new export potential as climate change affects production in other regions of the world. A broader set of solutions that should also be considered is discussed in the final section of the report.

KEY CLIMATE CHANGE ADAPTATION TECHNOLOGIES

Agricultural R&D institutions, both public and private, have been at the forefront of technology development to counter the impact of climate change. Some of these technologies are already being implemented on farmers' fields on a limited scale and some are in laboratory trials, while others are still in development. Most of the technologies simulated in this study are either already in use or have shown promise in laboratory trials.

The emerging climate change adaptation technologies can be classified into: 1) seed varietal technologies, based on enhanced seed characteristics that counter the adverse effects of climate change; 2) soil fertility management technologies that promote sustainable management of soils that are degrading as a result of climate change; 3) irrigation water management technologies that increase efficiency in the application and utilization of irrigation water and complement soil fertility management; and 4) crop protection technologies, including new and innovative ways of protecting crops from existing and emerging regimes of weeds, diseases, and insects induced by changing climate. See Box 1 for more details and the brief review of recent literature on climate change adaptation technologies below.

Improved seeds

Salt-tolerant, drought-tolerant, and water-saving wheat and rice varieties have already played a significant role in raising yields on Egypt's farms, particularly for wheat (Abdelmageed et al. 2019; Kassim et al. 2018; Eid et al. 2007). Field trials continue to test the performance of pest- and drought-resistant seed varieties, some with promising results – for example, a high-yielding and weed-resistant variety of fava bean increased yields in the Sharkia governorate by 23–38 percent (ICARDA 2016a).

Irrigation technology and sources

Flood irrigation is still the predominant irrigation method in Egypt, though it is less efficient than drip or sprinkler technology, which can reduce evaporation losses while still meeting crop water requirements and improving yield quantity and quality (Bargout and Fraser 2018). Center-pivot sprinkler irrigation systems are generally used in the sandy soils of the "new lands" (agricultural area reclaimed from desert land) to reduce water wastage and address other agronomic challenges.

Agricultural drainage water is often reused for irrigation, particularly on salt-tolerant crops such as olives and dates. Municipal wastewater is also increasingly recycled for irrigation purposes (Kassim et al. 2018), as reflected in Table 1.2.

As a result of increasingly insufficient surface water sources, Egypt is relying more and more on both renewable and nonrenewable groundwater sources. Pumping this water requires substantial energy that can be increasingly provided by solar power. The use of nonrenewable groundwater mines water resources that may render them unavailable for future generations, and so should be considered with great care.

Water productivity techniques

Many water management techniques focused on improving water productivity can be employed for climate change adaptation at the farm level. For example, raised-bed planting for wheat shows 30 percent yield increases and a 25 percent reduction in water use (ICARDA 2016b). Protected or shaded production of vegetable crops can save water, boost yields, widen the planting window, and reduce temperature stress (Abou Hadid 2013).

Deficit-irrigation, which reduces water application to crops based on careful yield response to water assessment, while minimizing economic losses can be cost-effective for some crops. For example, Abo El-Enein et al. (2011) note that a deficit-irrigation strategy can be cost-effective for cotton, since application of 30 percent less water reduces yields by only 10 percent. Changing sowing dates to plant earlier during the cooler months is another adaptation method already practiced by many farmers.

Optimum use of fertilizer

A recent study of smallholder farmers in Egypt shows that nitrogen fertilizers are applied in excess of government recommendations by farmers with access to subsidized fertilizers. Excessive application of nitrogen fertilizers results in only marginal increases in productivity but increases the potential adverse effects of soil acidification and water pollution (Kurdi et al. 2020).

CLIMATE CHANGE ADAPTATION STRATEGIES

The future of Egypt's agriculture in the face of climate change ultimately depends on the extent of its impacts and the policies and investment choices the government makes, as well as the possibilities for agricultural producers and consumers to adapt to these impacts and to take up solutions.

Historically, Egypt has achieved some of the highest agricultural productivity levels in the world. Moreover, Egypt's commercial farmers are receptive to modern technologies, including advanced seed and irrigation technologies. At the same time, land and water scarcity is increasingly constraining agricultural growth, with slowing or even negative yield growth for key commodities over the last 10 years. With climate change as a growing and permanent threat, it is important that Egypt not only consider but also rapidly advance on key climate change adaptation strategies in the agriculture sector. To inform these strategies, we simulate a series of adaptation scenarios aimed at maintaining agricultural productivity and supporting the country's food security objective. We recommend that these technologies and approaches cover all agricultural commodities and not be limited to strategic or food security crops; that the focus be on suites of technologies for larger impact, rather than on "silver bullet" or single-technology strategies; and that Egypt welcome and support adaptation strategies in other countries, as this ultimately benefits Egyptian consumers (see below on global cooperation efforts).

Agriculture-wide approach

Advances in developing modern production technologies have historically been biased toward cereals and staples in order to address caloric insufficiencies and basic energy needs to ensure survival. As the threat of climate change affects all food crops, recent agricultural R&D thrusts are becoming much broader and include minor and nonstaple crops. Farmers producing all types of agricultural commodities require technical support from the research community as well as an enabling environment from the government to cope with climate change.

Within the context of the methodology used in this study, the impact on productivity and welfare gains are highest when adaptation technologies are developed and scaled up for all agricultural commodities affected by climate change. To ensure that adaptation measures support all agricultural commodities, national agricultural research and extension systems must expand activities for secondary or nonstrategic crops, or gain needed insights through exchange with other research systems and incorporate these into national adaptation and mitigation strategies.

Technology suite approach

Socioeconomic analyses of agricultural technologies are most often undertaken ex post to evaluate the effectiveness and performance of the technology. However, as technologies move from laboratories to farmers' fields to the community and to the entire country, their environmental and socioeconomic variabilities also increase; and the economic viability and social acceptability become as important, if not more so, as the biophysical suitability of the technology. Ultimately, no single technology is best when scaled up to the governorate and national levels – rather, a selection or package of technologies is more appropriate. This can be called the "suite of technologies" approach. Under the scenarios we examine, each farmer can choose the most appropriate technology set from one or various groups of technologies, reflecting his or her specific biophysical and socioeconomic conditions. Technologies are grouped according to their key characteristics and, as feasible, according to whether they address climate change directly or indirectly (Box 1).

SIMULATION SET-UP AND MODEL IMPLEMENTATION

This study used IFPRI's IMPACT model, with a scenario-based setup to explore the effectiveness of different technology suites in countering the impact of climate change on the agriculture sector. Indicators of productivity (yield and production), food security (consumption and trade), and welfare (consumer and producer surpluses) were used as metrics of effectiveness.

Adaptation policy scenarios

The different adaptation scenarios are summarized in Box 1 and described in more detail below.

Scenario 1. Seed varietal technology suite

Focuses on the development of new varieties with specific traits aimed at countering one or several climate change effects, such as tolerance to heat, drought, submergence, and salinity. These technologies are preventive in nature – that is, developed to minimize or prevent the adverse effects of heat, drought, flooding, and salinity on crop yields but not designed for additional yield increases. They focus on preventing the biophysical yield losses presented in Table 2.2. (Resistance to plant diseases induced by climate change, such as bacterial blight, is considered in Scenario 4 as a novel approach to crop protection.)

Scenario 2. Soil fertility management technology suite

Focuses on improved farm management practices that enhance, retain, or restore soil fertility in order to sustainably improve soil quality and productivity. These practices affect the negative productivity impacts of climate change indirectly through higher soil fertility. They include no-till and direct seeding; integrated soil fertility management; organic farming; brown and green manuring; and precision-agricultural techniques, such as temperature and soil moisture sensors. As simulated in this study, on average, these practices increase yields of rice by 5 percent, maize by 16 percent, and wheat by 9 percent.

Scenario 3. Irrigation water management technology suite
Designed to increase irrigation efficiency, minimize water losses, increase water availability, and optimize water productivity and yields. These technologies serve to compensate for lack of rainfall and to alleviate the temperature and water stresses induced by climate change. The suite includes water harvesting, laser land leveling, alternate wet and dry (AWD) irrigation, and precision water application. In the Nile Delta, this also includes continued application of gravity irrigation and drainage infrastructure designed to flush accumulated salts from the soil and to push back saltwater intrusion. They work both as preventive and productivity-enhancing strategies. As simulated, they minimize water-stress and often reduce crop water requirements.

Scenario 4. Crop protection technology suite
Focuses on the control of new generations of weeds, pests, and diseases – recognizing that changing rainfall patterns and temperatures alter the environment and ecosystems and thus engender new pest and disease regimes. These technologies are being developed in response to potential geographic migration of pests and diseases and mutations to survive the changing climatic environment. Considering the uncertainty around pests and diseases, we assume that these technologies achieve 5 percent yield-damage prevention on average.

Scenario 5. Stacked technology suite
Combines complementary technologies to achieve additional productivity increases and cost savings. Technologies from different technology groups tend to be more complementary than technologies within the same groups, which can be mutually exclusive (an exception would be combined heat and drought stress tolerance). The joint application of complementary technologies – also called technology stacking – combines complementary technologies from each group, that is, seed technology, soil fertility, water management, and crop protection. As simulated, stacked technologies can have an increased productivity effect of 23 percent for rice, 52 percent for maize, and 17 percent for wheat over 20 years (2020–2040). But due to the complexity and costs involved in stacking these technologies, projected peak adoption is only 40 percent for the moderate assumption, which effectively reduces productivity gains at the national level to 9 percent for rice, 21 percent for maize, and 7 percent for wheat.

Baseline climate scenarios. A *no-climate-change* (No CC) scenario, which assumes continuation of historical (1970–2000) climate and productivity growth based on business-as-usual levels of investment in R&D, serves as the baseline climate (see Appendix Table A3 for yield growth under this scenario). The three GCMs (GFDL, HGEM, and IPSL) serve as *with-climate-change* scenarios (CC, with-CC, or average CC). All five adaptation scenarios above were simulated under climate change for each of the three GCMs.

Rate of scaling up - moderate and high adoption. The five adaptation technology scenarios were simulated under the climate change scenarios with two levels of on-farm adoption: a moderate level assuming adoption of 60 percent for Scenarios 1–4 and 40 percent for Scenario 5; and a high level assuming 80 percent adoption for Scenarios 1–4 and 60 percent for Scenario 5. All simulations run from 2020 to 2050 and assume that peak

BOX 1 DESCRIPTION OF SELECTED CLIMATE-RESILIENT TECHNOLOGIES

Adaptation scenarios	Individual technology	Brief description
Scenario 1 Seed Varietal Technology Suite	Heat tolerance	Improved varieties that allow plants to maintain yields at higher temperatures.
	Flood tolerance	Improved varieties that are tolerant to flooding or heavy rainfall conditions. Some varieties can withstand excess water and even prolonged underwater submergence.
	Drought tolerance	Improved varieties with enhanced soil moisture uptake capabilities and reduced vulnerability to water deficiency that allow the plants to have better yields than regular varieties.
	Salinity tolerance	Improved varieties that are more resistant to saline water, including saltwater intrusion and rising sea level due to changing climate.
Scenario 2 Soil Fertility Management Technology Suite	No-till and direct seeding	Minimum or no soil disturbance, often in combination with residue retention, crop rotation, and use of cover crops that retain or enhance natural soil fertility. Direct seed broadcasting is also often used in this technology.
	Integrated soil fertility management	Combination of chemical fertilizers, crop residues, and manure/compost based on site-specific soil fertilizer management, soil analysis, and nutrient-deficiency fertilizer matching.
	Organic farming, brown and green manuring	Use of only organic sources of fertilizers and pest and weed controls. No use of chemicals in crop production and crop protection.
	Full and partial intensification	Increase in the use of inputs to increase and optimize yield level. Usually involves increased fertilizer usage and land spaces to maximize light absorption.
	Precision agriculture	GPS-assisted delivery of agricultural inputs, as well as low-tech agricultural practices that aim to optimize management of crops (this includes effective plant spacing and use of appropriate planting windows). Localized and spot water, chemical, and fertilizer application.
Scenario 3 Irrigation Water Management Technology Suite	Water harvesting	Channeling water toward crop fields through macro- or micro-catchment systems or by using earth dams, ridges, or graded contours.
	Laser land leveling	Use of precision laser technology in the construction of bunds and land preparation to efficiently manage water flow and the application of irrigation water.
	Alternate wet and dry system	Water-saving technology that involves efficient application of irrigation water. Includes timing of irrigation to coincide with plant water demand at different stages of plant development, in combination with fertilizer application and weed control.
	Precision water application	Like precision agriculture but limited to site-specific and time-specific application and efficient non-application of irrigation water. Included also are the types of water delivery like drip and sprinkler.
Scenario 4 New Crop Protection Technology Suite	Weed protection	Chemical treatment to protect crops against existing weeds and changing weed regimes brought about by changing climate.
	Insect protection	Chemical treatment to protect crops against current and future onset of insect and arthropod pests.
	Disease protection	Chemical treatment to protect crops against diseases and pathogens related to changing climate.
Scenario 5 Stacked Technology Suite	Mix of complementary technologies	Combined application in the same farm of two or more complementary technologies from different technology suites. This specifically involves the application of three complementary technologies – one from seed varietal suite; another one from either soil-fertility or irrigation water suite; and one from the crop protection suite.

Source: Based on Rosegrant et al. (2014).

TABLE 3.1 YIELD EFFECTS OF MODERATE ADOPTION OF SUITE OF CLIMATE-RESILIENT TECHNOLOGIES FOR MAJOR FOOD GROUPS, EGYPT, BY 2050 (MODERATE ADOPTION RATE)

Food commodities	Yield effects of climate change	Minimum yield change to counter climate change	Suite of climate-resilient technologies				
			Seed technology	Soil fertility mgt	Irrigation water mgt	Crop protection	Stacked technology
Yield	*percent change from No CC*		*percent change from climate change scenarios*				
All food crops	-6.2	6.6	9.9	4.7	1.7	3.7	13.2
All cereals	-10.4	11.6	7.2	7.3	1.5	3.1	12.7
Maize	-19.5	24.3	13.6	10.1	1.4	3.0	19.0
Rice	-8.5	9.3	7.3	3.5	1.7	3.5	10.4
Wheat	-0.6	0.6	1.4	5.9	1.6	3.4	7.8
Fruits and vegetables	-8.3	9.0	12.4	6.0	2.0	4.6	16.5
Oil crops	-12.1	13.7	8.8	4.7	1.6	3.3	12.2
Pulses	-10.0	11.1	5.8	3.0	1.6	3.6	9.0
Roots and tubers	3.6	-3.4	4.3	2.3	1.6	2.3	6.6
Sugar crops	-13.3	15.3	8.5	4.3	1.6	3.3	11.6

Source: IMPACT simulations.

Notes: Boxed values mean that changes in yields from climate-resilient technologies fully counteract the effects of climate change. Values are averages of 3 GCMs: GFDL = General Fluid Dynamics Laboratory; HGEM = Hadley Centre Global Environmental Model; IPSL = Institut Pierre-Simon Laplace; all RCP 8.5 and SSP2.

adoption rates (that is, 60 percent for moderate and 80 percent for high adoption) and full damage prevention or productivity increases are achieved by farmer-adopters by 2040, starting from zero adoption in 2020. Note that this upper limit on adoption, though realistic, also introduces boundaries on the effectiveness of the technologies at the national level.

Who adapts? We simulate a scenario where only Egypt accelerates adaptation investments dramatically beginning in 2020 and up to 2040, and an alternative cooperation scenario where the rest of the world accelerates adaptation at the same pace. A comparison of these two scenarios allows us to assess the differences in benefits accruing to Egyptian consumers and producers from sole versus collaborative adaptation acceleration. As investments in adaptation efforts in the agriculture sector are low globally, both these scenarios are "what-if" or exploratory.

RESULTS AND DISCUSSION

All adaptation scenarios presented in Box 1 were individually simulated using the three climate models (GDFL, HGEM, and IPSL) under SSP2 and RCP 8.5 assumptions at the two rates of adoption. Results for moderate adoption are presented and discussed in detail in here; simulation results for the aggressive adoption rate are presented in the Appendix Tables A6 to A8. All simulations were run for 2020 to 2050, with peak adoption and full impact of adaptation achieved by 2040. Data presented average the results across the three climate change scenarios.

Productivity effects

Our simulation results suggest that wheat and roots and tubers are least affected by climate change in Egypt, with simulated overall yield declines of 0.6 percent for wheat and an improvement of 3.6 percent for roots and tubers (Table 3.1). Intuitively, for these crops, any adaptation technology focused on improving soil health, water-use efficiency, addressing new pests and diseases, and improving varieties, should be effective in counteracting climate change. Fruits and vegetables, as a group, are next least affected by climate change, and improved varieties are projected to achieve substantial yield gains over a business-as-usual scenario. As expected, productivity effects are larger if climate-resilient varieties are combined with improved soil fertility, water management, and crop protection.

A similar pattern of results is obtained under the high adoption scenario for the four single technology groups. The stacked technologies are particularly effective for maize, oil crops, pulses, and sugar crops (Appendix Table A6). With limited area for expansion, the effects on production mirror yield results, except for sugar and oil crops under the stacked technology simulations, where combined area and production increases were larger than the yield benefits from adaptation strategies (Table 3.2 and Figure 3.2).

Note that unless substantial emissions reductions are achieved in the near future, no single technology group is expected to fully counter the multifaceted impacts of climate change at the national scale. Instead, several adaptation strategies likely need to be combined. Moreover, for those crops most adversely affected by climate change, including maize, oil crops, pulses, and sugar crops, even the proposed stacked adaptation strategies are insufficient to recoup climate-change induced yield losses (see Figure 3.1 for maize). Moreover, while adopters of the technology suites can regain some of the lost yield, even non-adopters can benefit from higher prices for agricultural commodities (Table 2.6).

Impacts on food production, food security, and trade

If Egypt accelerates adaptation efforts unilaterally, without concomitant global effort, Egyptian producers benefit from increased productivity. However, consumers see few gains, because the change in Egyptian agricultural production, on its own, does not significantly affect global prices of

FIGURE 3.1 MAIZE, RICE, AND WHEAT YIELDS UNDER ALTERNATIVE CLIMATE AND TECHNOLOGY SCENARIOS, EGYPT, BY 2050 (MODERATE RATE OF ADOPTION)

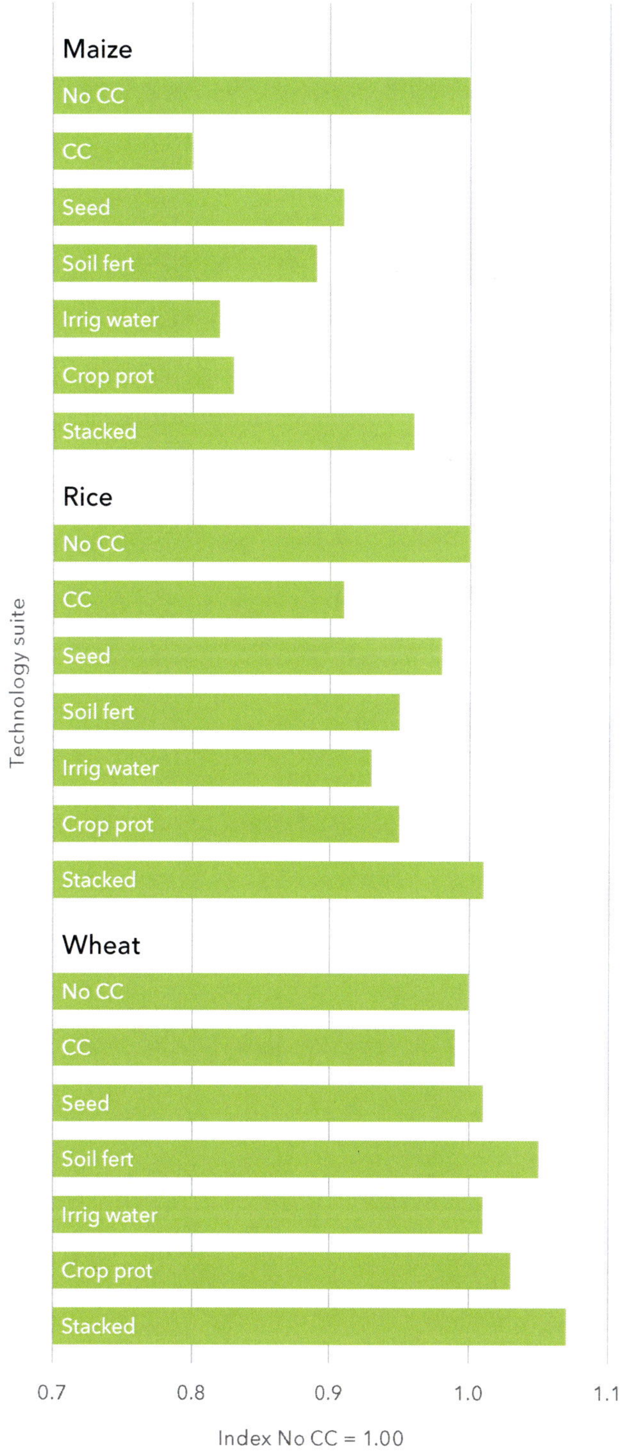

Source: IMPACT simulations.

TABLE 3.2 PRODUCTION EFFECTS OF ADOPTION OF A SUITE OF CLIMATE-RESILIENT TECHNOLOGIES FOR MAJOR FOOD GROUPS, EGYPT, BY 2050 (MODERATE ADOPTION RATE)

Food commodities	Production effects of climate change	Minimum production change to counter climate change	Suite of climate-resilient technologies				
			Seed technology	Soil fertility mgt	Irrigation water mgt	Crop protection	Stacked technology
Production	*percent change from No CC*		*percent change from climate change scenarios*				
All food crops	-5.7	6.1	10.5	5.1	1.8	3.9	14.1
All cereals	-11.9	13.5	7.3	8.6	1.7	3.4	13.9
Maize	-21.8	27.9	17.6	13.1	1.4	3.1	24.2
Rice	-6.4	6.9	9.1	3.7	2.3	5.0	12.4
Wheat	-2.3	2.3	0.5	6.7	1.7	3.8	6.7
Fruits and vegetables	-9.4	10.3	15.8	6.6	2.4	5.6	20.1
Oil crops	-5.7	6.0	8.0	3.3	1.2	2.8	10.2
Pulses	-23.9	31.4	5.9	1.6	2.0	4.7	8.9
Roots and tubers	3.1	-3.0	3.6	1.3	1.7	2.1	5.4
Sugar	-3.5	3.7	8.8	3.2	1.4	3.1	10.8

Source: IMPACT simulations.

Notes: Boxed values mean that changes in production from climate-resilient technologies fully counteract the effects of climate change. Values are averages of 3 GCMs: GFDL = General Fluid Dynamics Laboratory; HGEM = Hadley Centre Global Environmental Model; IPSL = Institut Pierre-Simon Laplace; all RCP 8.5 and SSP2.

FIGURE 3.2 IMPACT OF CLIMATE-RESILIENT TECHNOLOGIES ON PRODUCTION OF MAJOR FOOD CROPS, EGYPT, BY 2050 (MODERATE ADOPTION RATE)

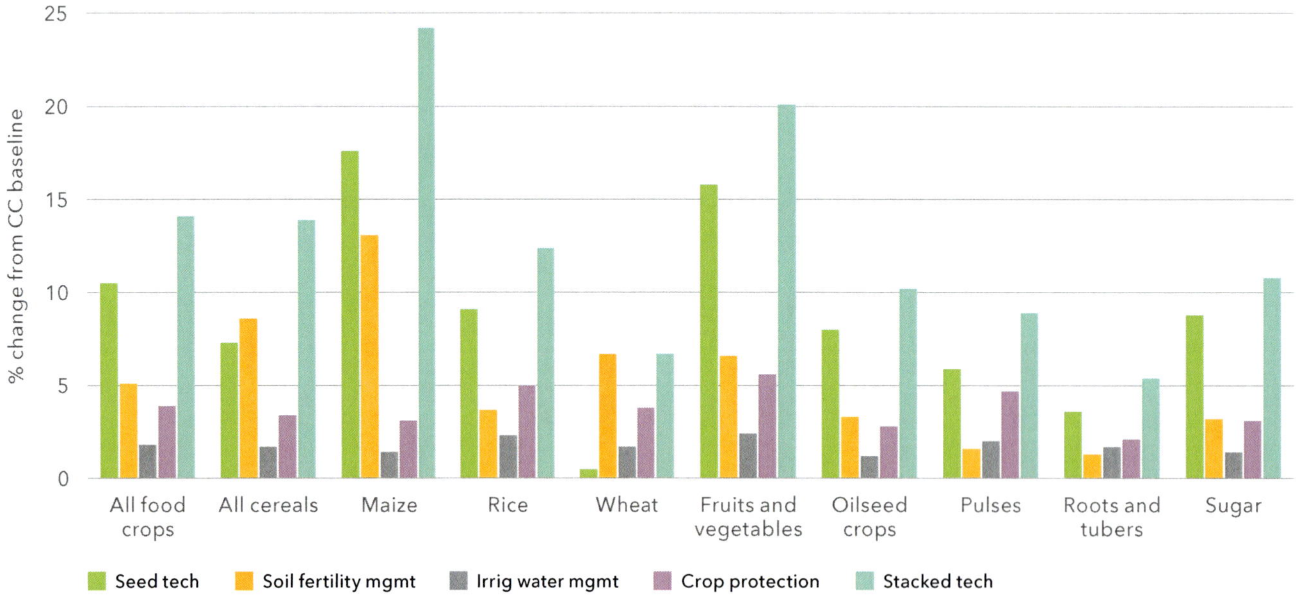

Source: IMPACT simulations.

TABLE 3.3 CHANGES IN NET TRADE STATUS UNDER CLIMATE CHANGE AND ADOPTION OF SUITE OF CLIMATE-RESILIENT TECHNOLOGIES, EGYPT, BY 2050 (MODERATE ADOPTION RATE, ONLY EGYPT ADAPTING)

				Suite of climate-resilient technologies				
Food commodities	No climate change	Climate change	Average of all technologies	Seed technology	Soil fertility mgt	Irrigation water mgt	Crop protection	Stacked technology
Net trade				'000 mt				
All food crops	-29,139	-27,810	-20,164	-16,840	-21,968	-25,833	-23,569	-12,609
All cereals	-35,957	-32,770	-31,116	-31,050	-30,737	-32,370	-31,955	-29,468
Maize	-17,609	-14,231	-13,092	-12,548	-12,976	-14,095	-13,930	-11,912
Rice	-2,868	-2,770	-2,603	-2,536	-2,675	-2,710	-2,642	-2,451
Wheat	-12,373	-12,558	-12,193	-12,721	-11,852	-12,373	-12,162	-11,855
Fruits and vegetables	12,387	7,777	13,055	16,014	11,226	9,023	10,723	18,287
Oilseed crops	-1,105	-1,025	-983	-958	-1,000	-1,015	-1,001	-941
Pulses	-1,653	-1,843	-1,819	-1,813	-1,835	-1,833	-1,819	-1,797
Roots and tubers	4,263	7,244	6,483	6,576	6,307	6,353	6,399	6,783
Sugar	-2,077	-2,076	-1,820	-1,663	-1,926	-2,012	-1,932	-1,568

Source: IMPACT simulations.

Note: Negative values mean net imports; positive values mean net exports. Values are averages of 3 GCMs: GFDL = General Fluid Dynamics Laboratory; HGEM = Hadley Centre Global Environmental Model; IPSL = Institut Pierre-Simon Laplace; all RCP 8.5 and SSP2.

agricultural commodities (Box 2). Thus, any unilateral adaptation effort would not improve domestic consumption, as prices are unaffected. But the increased national production can substitute for some of the increased import demand, such as for cereals, while net exports of fruits and vegetables would increase (Table 3.3). Every adaptation technology, on average, effectively contributes to reduced imports or increased exports. Food imports decline by 27 percent on average compared to a climate change scenario without adaptation, and by 31 percent compared to a scenario without climate change.

GLOBAL COOPERATION IN CLIMATE CHANGE ADAPTATION EFFORTS

For climate change *mitigation* efforts to succeed, global cooperation in reducing GHG emissions is needed, given the global nature and impact of climate change. Would *adaptation* to climate change work in the same way?

Results of the simulations in this study confirm that global cooperation in adaptation would result in large global welfare gains (Box 2). However, individual countries such as Egypt can still achieve substantial national welfare gains, whether or not the rest of the world cooperates. According to our calculations, the annual economic surplus generated if Egypt alone accelerates climate change adaptation in the agriculture sector is $0.73 billion per year; however, if the rest of the world also increases its focus on adaptation in the agriculture sector, Egypt's welfare gain would grow to $0.85 billion per year.

Adaptation to climate change undertaken with global cooperation is projected to substantially boost global production and lower food prices, compared with Egypt going it alone. In the case of global collaboration, prices of all food commodities decline, and by as much as 16 percent for maize and rice (Table 3.4). The average yield improvements from the simulated adaptation technologies are larger than the improvements needed to counteract climate change, and in several cases go beyond the yield improvements achieved under a no-climate-change scenario (Figure 3.3).

As a consequence of lower global prices, imports of food are more available and accessible to complement domestic production (Table 3.5). For Egypt, imports of cereals, including wheat, rice, and maize reach close to just 1 percent below no-climate-change levels. Only fruits and vegetables show continued export declines.

The implications of global efforts for food security and nutrition are more profound., Annual per capita consumption of foods and the daily intake of calorie equivalents are all within ±0.25 percent of levels without climate change (Table 3.6). Adaptation to climate change with global cooperation can reduce the disruption of global food markets, which is of particular importance for countries such as Egypt that are well integrated into these markets, and that import low-value, high-water-consuming cereals.

BOX 2 WELFARE IMPLICATIONS OF COOPERATION OR NON-COOPERATION BETWEEN EGYPT AND ROW ON CLIMATE CHANGE ADAPTATION EFFORTS, 2020-2050

Adaptation cooperation scenarios	Welfare measure		
	Producer surplus	Consumer surplus	Economic surplus
	net present value (billion US dollars)*		
Egypt adapts to climate change, ROW does not			
Rest of world, average	**-30.9**	**56.7**	**25.8**
Egypt, average	**21.1**	**0.9**	**22.0**
Seed Tech	31.0	1.3	32.3
Soil Fertility Mgt	13.9	0.6	14.5
Water Irrig Mgt	5.2	0.2	5.4
Crop Protection	10.8	0.5	11.3
Stacked Technologies	44.7	2.0	46.7
Annualized value, Egypt	*0.70*	*0.03*	*0.73*
Both Egypt and ROW adapt to climate change**			
ROW, average	**-1,419.0**	**2,852.8**	**1,433.9**
Egypt, average	**-20.6**	**46.1**	**25.5**
Seed Tech	-35.3	86.9	51.6
Soil Fertility Mgt	-19.5	48.3	28.8
Water Irrig Mgt	8.2	-30.3	-22.2
Crop Protection	-13.7	40.0	26.2
Stacked Technologies	-42.7	85.6	42.9
Annualized value, Egypt	*-0.69*	*1.54*	*0.85*

Source: Impact simulations.

Note: *Application of real discount rate equal to 3%. **Assumes the same scale (60%) of adaptation.

This table shows and emphasizes that by adapting to climate change welfare gains can be achieved by Egypt with or without the cooperation of ROW. However, Egyptian farmers would continue to gain more if ROW does not cooperate, while Egyptian consumers can benefit more if ROW cooperates in climate adaptation efforts.

TABLE 3.4 IMPACT OF CLIMATE CHANGE AND ADAPTATION STRATEGIES ON WORLD PRICES OF AGRICULTURAL COMMODITIES, BY 2050 (MODERATE ADOPTION RATE FOR BOTH EGYPT AND ROW)

Food commodities	Price effects of climate change	Minimum price decline to counter effects of CC	Averages of all technologies	Suite of climate-resilient technologies				
				Seed technology	Soil fertility mgt	Irrigation and water mgt	Crop protection	Stacked technology
World prices	*percent change from No CC*		*percent change from CC*					
All meat products	5.0	-4.7	-5.6	-5.90	-3.55	-4.60	-2.62	-11.26
Beef/Buffalo meat	2.7	-2.7	-3.6	-3.84	-2.48	-2.59	-1.78	-7.29
Mutton/Goat meat	1.5	-1.5	-2.0	-2.18	-1.36	-1.47	-0.98	-4.15
Poultry meat	8.7	-8.0	-9.0	-9.46	-5.55	-7.76	-4.17	-17.83
Dairy products	1.1	-1.1	-1.5	-1.62	-1.04	-1.07	-0.75	-3.09
Eggs	5.7	-5.4	-6.1	-6.43	-3.74	-5.25	-2.80	-12.31
All cereals	14.0	-12.3	-12.6	-11.29	-8.54	-11.43	-5.77	-25.97
Maize	22.8	-18.6	-15.9	-14.65	-9.59	-17.28	-5.62	-32.34
Rice, milled	18.7	-15.8	-15.1	-14.16	-7.39	-15.88	-6.91	-30.97
Wheat	2.5	-2.4	-7.5	-6.17	-9.76	-0.74	-6.26	-14.63
Fruits and vegetables	10.2	-9.2	-11.6	-13.47	-6.89	-8.48	-5.50	-23.51
Oil crops	20.0	-16.6	-15.2	-15.37	-7.87	-15.16	-6.42	-30.96
Pulses	7.5	-7.0	-9.7	-10.74	-5.60	-6.53	-5.97	-19.74
Roots and tubers	15.8	-13.7	-13.2	-12.24	-6.26	-13.98	-5.97	-27.38
Sugars	7.4	-6.9	-5.1	-5.68	-2.94	-4.45	-2.31	-10.28

Source: IMPACT simulations.

Note: Boxed values mean that changes in prices from climate-resilient technologies fully counteract the effects of climate change. Values are averages of 3 GCMs: GFDL = General Fluid Dynamics Laboratory; HGEM = Hadley Centre Global Environmental Model; IPSL = Institut Pierre-Simon Laplace; all RCP 8.5 and SSP2. Global cooperation assumes same 60 percent rate of adoption of climate resilient technologies for Egypt and ROW.

FIGURE 3.3 IMPACT ON WORLD PRICES OF MAJOR FOOD GROUPS OF CLIMATE CHANGE AND SUITE OF CLIMATE-RESILIENT TECHNOLOGY COUNTERMEASURES, BY 2050 (MODERATE ADOPTION RATE)

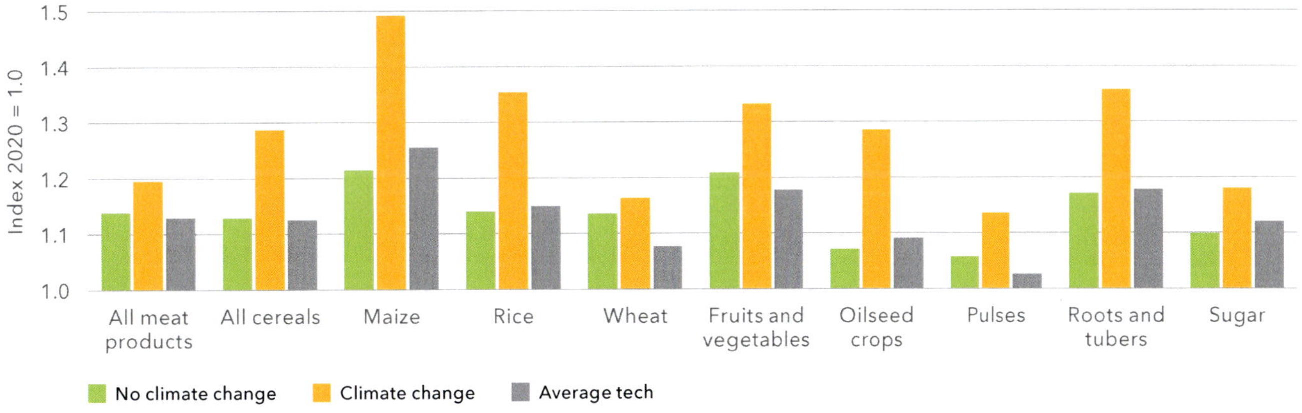

Source: IMPACT simulations.

TABLE 3.5 CHANGES IN NET TRADE OF MAJOR FOOD CROPS DUE TO CLIMATE CHANGE AND ADAPTATION STRATEGIES, EGYPT AND REST OF THE WORLD, BY 2050 (MODERATE ADOPTION RATE)

Food commodities	No climate change	Climate change	Egypt adapts, ROW does not	Egypt and ROW adapt
Net trade		'000 mt		
All food crops	-29,139	-27,810	-20,164	-32,694
All cereals	-35,957	-32,770	-31,116	-35,695
Maize	-17,609	-14,231	-13,092	-17,257
Rice	-2,868	-2,770	-2,603	-3,076
Wheat	-12,373	-12,558	-12,193	-12,209
Fruits and vegetables	12,387	7,777	13,055	6,668
Oilseed crops	-1,105	-1,025	-983	-1,143
Pulses	-1,653	-1,843	-1,819	-1,791
Roots and tubers	4,263	7,244	6,483	5,511
Sugar	-2,077	-2,076	-1,820	-2,137

Source: IMPACT simulations.

Note: Values are averages of 3 GCMs: GFDL = General Fluid Dynamics Laboratory; GEM = Hadley Centre Global Environmental Model; IPSL = Institut Pierre-Simon Laplace; all RCP 8.5 and SSP2. Negative values mean net imports; positive values mean net exports.

TABLE 3.6 CHANGES IN PER CAPITA FOOD INTAKE AND CALORIE CONSUMPTION DUE TO CLIMATE CHANGE AND ADOPTION OF CLIMATE-RESILIENT TECHNOLOGIES, EGYPT, BY 2050 (MODERATE ADOPTION RATE FOR BOTH EGYPT AND ROW)

Food commodities	Consumption				
	No climate change	With climate change		Average of climate-resilient tech	
	kg/year	kg/year	percent change from No CC	kg/year	percent change from No CC
All meat products	40.5	40.2	-0.75	40.6	0.24
Beef	21.8	21.7	-0.15	21.8	0.23
Lamb	1.8	1.8	-0.32	1.8	0.29
Poultry	17.0	16.7	-1.57	17.0	0.24
Dairy	45.9	45.9	-0.15	46.0	0.16
Eggs	4.1	4.0	-0.80	4.1	0.16
All food crops	684.5	670.2	-2.09	686.4	0.27
All cereals	235.2	231.7	-1.52	236.8	0.66
Maize	55.4	54.1	-2.32	55.2	-0.36
Rice	35.6	33.7	-5.38	35.6	-0.07
Wheat	140.9	140.6	-0.23	142.7	1.28
Fruits and vegetables	324.4	318.7	-1.74	325.8	0.45
Oilseed crops	9.1	8.5	-6.57	9.1	0.19
Pulses	13.9	14.1	1.57	14.1	0.98
Roots and tubers	45.3	41.8	-7.66	44.4	-2.07
Sugar	49.4	48.1	-2.48	49.0	-0.77
	kcal/day	kcal/day	percent change	kcal/day	percent change
Daily calorie consumption	3,789	3,723	-1.74	3,798	2.03

Source: IMPACT simulations.

Note: Values are averages of 3 GCMs: GFDL = General Fluid Dynamics Laboratory; GEM = Hadley Centre Global Environmental Model; IPSL = Institut Pierre-Simon Laplace; all RCP 8.5 and SSP2.

4. CONCLUSIONS AND POLICY IMPLICATIONS: INVESTMENTS FOR SUSTAINABLE AND RESILIENT AGRICULTURE

Based on the work described in this report, we offer a set of key findings and messages for the sustainable development of Egyptian agriculture and increased resiliency in the face of growing climate change impacts.

THE AGRICULTURE SECTOR AND CLIMATE CHANGE

Egypt has a vibrant agriculture sector

Agriculture and food production growth has been rapid over the last three decades, though growth has slowed recently. In aggregate, meat production grew at an annual rate of 3.8 percent from 1990 to 2020 and cereals by 1.9 percent. Export crops – fruits, vegetables, and roots and tubers (mainly potatoes) –grew by 3.5, 2.1, and 4.3 percent annually, respectively. The resilience of the agriculture sector to shocks has been proven during the recent global health crisis; agriculture has performed relatively better than the service and industry sectors.

Climate change is a lasting threat to agriculture

Changing temperature, rainfall, and PET. Egypt's desertic climate will see yet higher temperatures and even less rain, resulting in higher postential evapotranspiration (PET) and crop water demand. As projected by different climate models, annual temperature increases for Egypt are as large as 3.1°C for HGEM and highest for the July-August-September period. Annual rainfall is projected to decline, with the largest reduction of 15 mm (under the IPSL model), while PET is expected to increase by 5-21 percent.

Declines of yields and production. Due to climate change, food crop yields in Egypt are projected to decline by 10 percent, on average, between 2020 and 2050, reflecting heat stress (−4.9 percent), water stress (−4.1 percent), and increased salinity (−1.6 percent), with the largest impacts projected for maize (−16.2 percent), sugar crops (−12.0 percent), and fruits and vegetables (−11.7 percent). When climate-change-induced changes in food supply and demand are considered, yield declines can further worsen for cereals, especially for maize, sugar, and pulses. Combined declines result in a projected reduction of food crops of by 5.9 percent, cereals by 11.9 percent (including 21.8 percent for maize), rice by 6.4 percent, and wheat by 2.3 percent, as well as fruits and vegetables by 9.4 percent.

Increases in food prices. Global declines in food production, in turn, are projected to drive up prices of food by as much as 22.8 percent for maize and by 18.7 percent for rice. Prices of oil crops, root and tuber crops, and poultry are also projected to increase significantly, and greatly affecting food security for poorer segments of Egypt's population. Higher global prices will reduce Egypt's food import demand, and also dampen demand for Egypt's exports. The resulting tighter food markets will make it more difficult for Egypt to rely on food imports to augment domestic supplies.

The economic costs of climate change are substantial

Within the agriculture sector alone, the economic costs to the Egyptian population are estimated at $55.3 billion for the 30-year period (2020-2050), or $1.84 billion per year. Egyptian consumers bear most of the costs, at $2.21 billion per year, while producers gain an average of $0.37 billion per year, implying that higher prices more than compensate for productivity losses. However, most farmers – especially smallholder farmers – are marginal producers and net buyers of food, and thus are expected to suffer net economic losses from the combined producer and consumer effects.

Impacts of climate change are exacerbated by land and water scarcity

The scarcity of fertile land and the high rate of urbanization further exacerbate the production impacts of climate change. About 75,000 ha of fertile land are estimated to have been lost to urban expansion in the last 20 years.

Despite growing water demands for domestic and industrial use, agriculture is projected to remain by far the largest water user. With an increasing population, growing investment in industry, and increasing water loss to evaporation, reduced water availability will be a growing threat to the country's agricultural economy.

Adaptation to climate change improves the resiliency and sustainability of agriculture

The expected impacts and costs of climate change for the agriculture sector make a compelling case for adaptation, as well as for Egypt's participation in global climate change mitigation efforts.

This report has focused on investment in agricultural R&D to develop and implement technologies as a direct response and countermeasure to climate change. Several technology suites were identified and simulated – all aim to prevent, minimize, or mitigate the impact of climate on the agriculture sector.

An additional set of adaptation strategies that also directly or indirectly counter the impact of climate change are presented below to highlight the larger spectrum of adaptation options beyond technologies. Further study should assess these options with analytical rigor, as the technological options were assessed here.

KEY ADAPTATION RESPONSES IN TECHNOLOGY AND INFRASTRUCTURE

The direct effects of climate change on yields and production are biophysical, through changes in temperature and rainfall patterns and their effects on land and water quality and quantity.

Direct countermeasures that target the biophysical impacts of climate change include: 1) investments in agriculture R&D to develop new technologies to counter or alter the effects of temperature (heat) and rainfall (droughts or flooding); and 2) investment in new land or water resources (infrastructure response).

Technical response – investment in agricultural R&D

Investment in agricultural R&D enhances the capability of the global research system to develop climate change adaptation technologies as well as of national systems to adapt technologies to local environmental and social conditions. Investment in R&D systems also includes strengthening national and local extension agencies in the promotion and dissemination of these technologies.

The technology suites include new/improved seed varieties with traits focused on combating climate change, soil fertility management practices, improved irrigation management, and enhanced crop protection. As many of these technologies work best when applied in tandem, stacking of complementary technologies and practices was also considered.

For crops less impacted by climate change in Egypt, such as fruits and vegetables, potatoes, rice, and wheat, increased investments in climate-change-responsive crop traits or a combination of measures including investments in soil fertility improvement, water management, and crop protection can counteract the adverse impacts of climate change. Across food crops, investments in climate-resilient seed technologies provide the largest returns, followed by investments in soil fertility, crop protection, and irrigation.

But for those crops hit hardest by climate change in Egypt, including maize, oil crops, pulses, and sugar, no single technology group can fully counter the multifaceted impacts of climate change at the national scale, and even stacking of technologies will not return productivity to no-climate-change levels. Therefore, advances on a larger set of activities are needed, as almost all crops require multiple improvements and specific on-farm adjustments. At this point, there is little information on the extent to which Egyptian farmers have adopted climate change adaptation strategies or what these strategies are. Additional field-level work can help shed light on farmer-preferred adaptation strategies and aspirations.

Given the considerable economic costs to the Egyptian population from climate change impacts on agriculture, estimated here at $1.84 billion per year, and given that technology-focused strategies hold great promise for adaptation, consideration should be given to whether these technologies are also cost effective. While their cost was not addressed here, several are already under development and they appear to be cost-effective both in terms of reducing the country's net import bill and in reducing the risk of growing food insecurity, which is heightened if other countries do not also accelerate adaptation investments.

With its strong linkages with world food markets, Egypt will also benefit from technology-driven adaptation efforts elsewhere in the world. Global cooperation on adaptation will mean lower food prices and less pressure on consumers. Under this scenario, only fruits and vegetables show continued export declines.

Thus, Egypt will benefit greatly from accelerating adaptation efforts in-country. It should focus on those adaptation options that do not further increase GHG emissions. Egypt should also actively support adaptation efforts elsewhere, given mutual benefits from lower food prices and the associated reduced food import bill.

Infrastructure response – investment in new lands and water sources

Fertile land is fast becoming scarce, due to very limited fertile lands along the Nile suitable for agriculture and continued loss to urban expansion. Over the past 40 years, the Egyptian government has been converting nonfertile desert lands into "new lands" for agriculture. Plans for reclamation are to continue at an annual rate of 63,000 ha, for a total of 588,000 ha by 2030, according to the country's Sustainable Agricultural Development Strategy of 2030 (SADS 2030) (MoALR 2020).

Under the SADS 2030 approach, treated wastewater will be used to irrigate the new lands. In the meantime, groundwater abstractions from the Nubian Sandstone Aquifer System are the main source of water. Nile River-linked aquifers are a further source. The new lands constitute about 15 percent of total irrigated cultivated areas, largely using drip and sprinkler irrigation technologies (McCarl et al. 2015; Viscomi 2010).

Developing these new lands may be less cost effective than using new seed technologies and enhanced agronomic practices on existing irrigated lands. In addition, there are concerns about the environmental impacts and sustainability of using nonrenewable groundwater sources.

COMPLEMENTARY ADAPTATION INVESTMENTS

Additional, indirect measures that can complement direct agricultural interventions are summarized here. These were not covered by the study but may benefit from further analysis.

Market response – market-based allocation of resources

Removal of price support and input subsidies. Prior to the reform programs that began in the late 1980s, crop area allocations, crop rotation schedules, and production quotas at fixed prices below the international market price were all government-mandated, with corresponding subsidized consumer prices. The Egyptian government also fixed certain output prices (for example, wheat prices) and input prices (for example, fertilizers and fertilizer raw materials), which contributed to suboptimal input use (Kassim et al. 2018; Kurdi et al. 2020). The reform programs of the 1990s reduced subsidies on agricultural inputs, liberalized markets, removed controls on prices and marketing restrictions for major crops, privatized public firms, and encouraged private sector participation in agriculture and agricultural trading (Baffes and Gautam 1996; Cassing et al. 2009; Gouda 2016).

Changing food self-sufficiency goals to food accessibility goals. Despite this liberalization, agricultural policy in Egypt continues to view food self-sufficiency as an indicator of national food security. In its revised strategic development plans for 2015–2030, the government aims to increase wheat and maize production to reach self-sufficiency in wheat and maize by 2030. Rather than food self-sufficiency, some proponents of market-incentive approaches suggest that increasing food accessibility – for example, through increased purchasing power, diversification in consumption and production, and access to trade – would better support multiple national development goals, including increased climate resilience.

Water pricing. Pricing of water for irrigation is a contentious topic everywhere in the world, and Egypt is no exception. Unpriced water use in agriculture is equivalent to input subsidies that result in suboptimal uses. Water pricing reforms can take many forms, including paying farmers to use less water or supporting farmers to trade irrigation water within specific canals. More research is needed on farmer preferences and effectiveness of alternative options.

Trade response – trading in virtual water, comparative advantage, and high-value crops

Egypt's food exports account for about 20 percent of total exported goods, boosted by a 12-fold increase in exports of high-value fruits and vegetables over the last 10 years. Imports of lower-value cereals and pulses similarly increased, accounting for 21 percent of total imports. Trade in virtual water – that is, water used in food production that is "embedded" in the traded products – is a means for water-scarce countries to access water-intensive food supplies from better water-endowed countries. In limited ways, Egypt is already trading in virtual water by restricting production of some crops, such as rice, and by promoting imports of water-intensive crops (rice and wheat). Egypt does not restrict exports generally; however, exports of some strategic crops have been prohibited in the past (for example, rice in 2008–2014). As such, Egypt has already adopted a strong trade-based water-savings and climate change adaptation policy.

The trade-based approach to food security is considered particularly suitable in the Middle East and North Africa region, as a means to free up water for the many non-irrigation water uses of the growing population. Moreover, Egypt is itself a competitive agricultural exporter, with a comparative advantage in agriculture and high yields of major agricultural crops (Breisinger et al. 2012; Nin-Pratt et al. 2018; Bargout and Fraser 2018; Kurdi et al. 2020).

Economic growth response

Sustainable agricultural production practices are an excellent foundation for climate-resilient agriculture. Egypt's SADS 2030 focuses on the sustainable use of natural resources; improving productivity; increasing market competitiveness; improving food security; increasing investment opportunities; and improving rural livelihoods. The implementation of this strategy should therefore also improve climate resilience in agriculture.

The agriculture sector's importance in the Egyptian economy is declining in terms of its relative share of both income and employment, as the service and industry sectors expand. This gradual change could lead to further specialization in high-value agricultural products and continued phasing out of water-intensive crops, which could further improve the climate resilience of Egypt's agriculture sector.

REFERENCES

Abdelmageed, K., X.H. Chang, D.M. Wang, Y.J. Wang, Y.S. Yang, G.C. Zhao, and Z.Q. Tao. 2019. "Evolution of Varieties and Development of Production Technology in Egypt Wheat: A Review." *Journal of Integrative Agriculture* 18 (3): 483–495. https://doi.org/10.1016/S2095-3119(18)62053-2

Abo El-Enein, R., M. Sherif, M. Karrou, T. Oweis, B. Benli, and H. Farahani. 2011. "Towards Sustainable and Improved Water Productivity in the Old Lands of the Nile Delta." In *Improving Water and Land Productivities in Irrigated Systems*. Community-Based Optimization of the Management of Scarce Water Resources in Agriculture in CWANA Report No. 10, eds. M. Karrou, T. Oweis, B. Benli, and A. Swelam, 9–29. Aleppo, Syria: ICARDA. https://repo.mel.cgiar.org/handle/20.500.11766/8447

Abou Hadid, A.F. 2013. "Protected Cultivation for Improving Water-Use Efficiency of Vegetable Crops in the NENA Region." In *Good Agricultural Practices for Greenhouse Vegetable Crops: Principles for Mediterranean Climate Areas*, eds. W. Baudoin, R. Nono-Womdim, N. Lutaladio, A. Hodder, N. Castilla, C. Leonardi, S. De Pascale, and M. Qaryouti, 137–149. Rome: FAO. http://www.fao.org/3/a-i3284e.pdf

Alissa, S. 2007. "The Political Economy of Reform in Egypt: Understanding the Role of Institutions." *Carnegie Papers* No. 5. Beirut: Carnegie Middle East Center, Carnegie Endowment for International Peace. https://carnegieendowment.org/files/cmec5_alissa_egypt_final.pdf

Alston, J.M., P.W. Norton, and P.G. Pardey. 1995. "Science under Scarcity: Principles and Practice for Agricultural Research Evaluation and Priority Setting." London, UK, and Ithaca, NY: Cornell University Press for the International Service for National Agricultural Research. https://www.cabi.org/bookshop/book/9780851992990/

Baffes, J., and M. Gautam. 1996. "Price Responsiveness, Efficiency, and the Impact of Structural Adjustment on Egyptian Crop Producers." *World Development* 24 (3): 765–771. https://doi.org/10.1016/0305-750X(95)00164-8

Bargout, R.N., and E.D.G. Fraser. 2018. "Addressing Water Scarcity in the Nile Delta: Virtual Water, Fresh Water, and Desalination." *Global Societies Journal* 6: 62–100. https://escholarship.org/uc/item/43q236m8

Breisinger, C., M. Raouf, M. Wiebelt, A. Kamaly, and M. Karara. 2020. *Impact of COVID-19 on the Egyptian Economy: Economic Sectors, Jobs, and Households*. MENA Policy Note 6. Washington, DC: IFPRI. https://doi.org/10.2499/p15738coll2.133764

Breisinger, C., O. Ecker, P. Al-Riffai, and B. Yu. 2012. *Beyond the Arab Awakening: Policies and Investments for Poverty Reduction and Food Security*. IFPRI Food Policy Report 25. Washington, DC: IFPRI. https://ebrary.ifpri.org/utils/getfile/collection/p15738coll2/id/126781/filename/126992.pdf

Cassing, J., S. Nassar, G. Siam, and H. Mousa. 2009. "Arab Republic of Egypt." In *Distortions to Agricultural Incentives in Asia*, eds. K. Anderson and W.A. Masters, 71–98. Washington, DC: World Bank. http://elibrary.worldbank.org.offcampus.lib.washington.edu/docserver/download/9780821376621.pdf

CAPMAS (Central Agency for Public Mobilization and Statistics). 2019. *Egypt in Figures: Water Resources*. https://www.capmas.gov.eg/Pages/Publications.aspx?page_id=5104&Year=23608

Eid, H.M, S.M. El-Marsafawy, and S.A. Ouda. 2007. "Assessing the Economic Impacts of Climate Change on Agriculture in Egypt: A Ricardian Approach." Policy Research Working Paper 4293. Washington, DC: World Bank. http://hdl.handle.net/10986/7485

FAO (Food and Agriculture Organization of the United Nations). 2020. FAOSTAT [database]. http://www.fao.org/faostat/en/#home

Gautam, H.R., M.L. Bhardwaj, and R. Kumar. 2013. "Climate Change and Its Impact on Plant Diseases." *Current Science* 105 (12): 1685–1691. http://www.jstor.org/stable/24099750

Gouda, D. 2016. *Social Capital and Local Water Management in Egypt*. Cairo: American University in Cairo Press.

Hammam, A.A., and E.S. Mohamed. 2020. "Mapping Soil Salinity in the East Nile Delta Using Several Methodological Approaches of Salinity Assessment." *Egyptian Journal of Remote Sensing and Space Sciences* 23 (2): 125–131. https://doi.org/10.1016/j.ejrs.2018.11.002

Ibrahim, F.N., and B. Ibrahim. 2003. *Egypt: An Economic Geography*. London and New York: I.B. Taurus

ICARDA (International Center for Agricultural Research in the Dry Areas). 2016a. "Enhancing Food and Nutrition Security Against Climate Change: Stronger Wheat-Legume Systems in West Asia and North Africa." In *ICARDA Annual Report 2015*. Amman. https://repo.mel.cgiar.org/handle/20.500.11766/7006

ICARDA. 2016b. "Raised-Bed Planting: Maximizing Water Use Efficiency." ICARDA News, Nov. 9. https://www.icarda.org/update/raised-bed-planting-maximizing-water-use-efficiency#sthash.6lrhsta8.oJZMRTvd.dpbs

IIASA (International Institute for Applied Systems Analysis). 2013. SSP Database. Accessed 2019. https://secure.iiasa.ac.at/web-apps/ene/SspDb/dsd?Action=htmlpage&page=about

IIASA. 2015. RCP Database. Accessed 2019. www.iiasa.ac.at

IPCC (Intergovernmental Panel on Climate Change). 2014. *Climate Change 2014: Synthesis Report. Contribution of Working Groups I, II and III to the Fifth Assessment Report of the Intergovernmental Panel on Climate Change*. Geneva.

Kassim, Y., M. Mahmoud, S. Kurdi, and C. Breisinger. 2018. "An Agricultural Policy Review of Egypt: First Steps Towards a New Strategy." IFPRI Middle East and North Africa Regional Program Working Paper 11. International Food Policy Research Institute, Washington, DC. http://ebrary.ifpri.org/cdm/ref/collection/p15738coll2/id/132788

Kotb, T.H.S., T. Watanabeb, Y. Oginob, K.K. Tanji. 2000. "Soil Salinization in the Nile Delta and Related Policy Issues in Egypt." *Agricultural Water Management* 43 (2): 239-261. https://doi.org/10.1016/S0378-3774(99)00052-9

Kurdi, S., M. Mahmoud, K.A. Abay, and C. Breisinger. 2020. "Too Much of a Good Thing? Evidence that Fertilizer Subsidies Lead to Overapplication in Egypt." IFPRI Middle East and North Africa Regional Program Working Paper 27. International Food Policy Research Institute, Washington, DC. https://doi.org/10.2499/p15738coll2.133652

MoALR (Ministry of Agriculture and Land Reclamation). 2020. *Egypt 2030 Updated Sustainable Agriculture Development Strategy (Executive Summary)*. Cairo.

McCarl, B.A., M. Musumba, J.B. Smith, et al. 2015. "Climate Change Vulnerability and Adaptation Strategies in Egypt's Agricultural Sector." *Mitigation and Adaptation Strategies for Global Change* 20 (7): 1097–1109. https://doi.org/10.1007/s11027-013-9520-9

Müller, C., and R.D. Robertson. 2014. "Projecting Future Crop Productivity for Global Economic Modeling." *Agricultural Economics* 45 (1): 37–50. http://dx.doi.org/10.1111/agec.12088

Nin-Pratt, A., H. El-Enbaby, J.L. Figueroa, H. ElDidi, and C. Breisinger. 2018. "*Agriculture and Economic Transformation in the Middle East and North Africa: A Review of the Past with Lessons for the Future*. Food Policy Report. Washington, DC: IFPRI; Rome: FAO. https://doi.org/10.2499/9780896292956

Omran, E-S.E. 2020. "Exploring Changes in the Agricultural Calendar as a Response to Climate Variability in Egypt." In *Climate Change Impacts on Agriculture and Food Security in Egypt*, Springer Water series, eds. E-S.E. Omran and A. Negm, 249–271. Cham, Switzerland: Springer. https://doi.org/10.1007/978-3-030-41629-4_12

Perez, N.D., and M.W. Rosegrant. 2018. "A Partial Equilibrium Approach to Modelling Alternative Agricultural Futures under Climate Change." In *The Future of Philippine Agriculture under a Changing Climate: Policies, Investments and Scenarios*, eds. M.W. Rosegrant and M.A. Sombilla, 450–491. Singapore: ISEAS-Yusof Ishak Institute.

Robinson, S., D. Mason d'Croz, S. Islam, T.B. Sulser, R.D. Robertson, T. Zhu, A. Gueneau, G. Pitois, and M.W. Rosegrant. 2015. "The International Model for Policy Analysis of Agricultural Commodities and Trade (IMPACT): Model Description for Version 3." IFPRI Discussion Paper 1483. International Food Policy Research Institute, Washington, DC. http://ebrary.ifpri.org/cdm/ref/collection/p15738coll2/id/129825

Rosegrant, M.W., J. Koo, N. Cenacchi, C. Ringler, R.D. Robertson, M. Fisher, C.M. Cox, K. Garrett, N.D. Perez, and P. Sabbagh. 2014. *Food Security in a World of Natural Resource Scarcity: The Role of Agricultural Technologies*. Washington, DC: IFPRI. http://dx.doi.org/10.2499/9780896298477

Sultan, N.A. 2020. "The Consistency of Export and Agricultural Policies in Egypt." Master's thesis, American University in Cairo, AUC Knowledge Fountain. https://fount.aucegypt.edu/cgi/viewcontent.cgi?article=1848&context=etds

Viscomi, J. 2010. *The Farthest Place: Social Boundaries in an Egyptian Desert Community*. Cairo Papers. Cairo: American University in Cairo Press.

World Bank. 2020. World Development Indicators [database]. Accessed at https://databank.worldbank.org/source/world-development-indicators

APPENDIX

A. ADDITIONAL TABLES: THE AGRICULTURE SECTOR AND THE REST OF THE WORLD

Tables A1 and A2 present the performance of Egypt's agriculture sector in terms of growth in area harvested and number of animals and productivity. These are a supplement to Table 1.2, where food production growth rates are presented in the main text.

Table A3 presents the yield growth rates for all the agricultural commodities simulated in the IMPACT model, under the business-as-usual scenario (no climate change) from 2020 to 2050. Table A4 presents the projected biophysical impacts of climate change on yields, together with the economic effects. Table A5 shows the projected world price of food by 2050.

TABLE A1 RECENT PERFORMANCE OF EGYPT'S AGRICULTURE SECTOR: Growth of area harvested and animal numbers, 1990–2020

Commodity/Indicators	Years				Annual growth rate by period				Change from 1990 to 2020
	1990	2000	2010	2020	1990-2000	2000-2010	2010-2020	1990-2020	
Area /Heads	*'000 ha/'000 heads/million birds*				*percent*				
Beef/Buffalo	2,235	2,998	2,631	2,339	3.20	**-1.13**	**-1.32**	**-0.06**	5
Sheep/Goats	3,864	4,161	4,799	2,924	0.20	1.46	**-4.17**	0.05	**-24**
Poultry	229	388	571	1,039	5.50	4.05	6.41	5.11	355
Layers	22	15	27	36	**-3.84**	5.93	3.11	2.53	63
Milk animals	4,017	5,864	6,359	6,160	3.55	0.68	**-0.41**	1.17	53
All cereals	2,103	2,704	3,119	2,967	2.19	1.72	**-0.39**	1.11	41
Maize	832	846	963	892	**-0.31**	0.63	**-0.39**	0.75	7
Rice	400	611	593	566	4.23	0.62	**-0.64**	0.41	41
Wheat	688	1,017	1,283	1,290	3.50	2.79	0.16	1.93	88
Fiber crops*	439	289	146	117	**-2.87**	**-5.68**	**-5.07**	**-4.79**	**-73**
Fruits	369	497	560	680	2.95	1.67	2.06	2.09	84
Nuts	2	4	8	6	4.95	5.90	**-3.55**	5.31	224
Vegetables	333	475	681	627	4.00	3.71	**-0.90**	2.87	88
Oil crops	530	434	312	326	**-1.15**	**-3.00**	**-0.78**	**-2.11**	**-39**
Pulses	191	168	116	79	**-1.05**	**-4.41**	**-3.03**	**-3.24**	**-59**
Roots and tubers	87	93	155	194	2.09	6.07	1.88	3.14	122
Sugar crops	130	180	252	349	3.18	3.34	3.46	4.04	187
Total area harvested	4,186	4,843	5,350	5,343	1.53	1.22	-0.08	0.89	28
Cropland area	2,600	3,345	3,634	3,870	2.73	0.85	0.63	1.13	49
Cropping intensity	1.61	1.45	1.47	1.38	-1.17	0.37	-0.71	-0.24	**-14**

Source: Constructed from FAOSTAT data.

Note: Values are 3-year moving averages of t-2 t-1 and t0. Latest data from FAO are for 2019 only, values for 2020 are trend-extrapolation of preceding 5 years of 2015–2019, before the 3-year moving averages were applied to 2020. *Though not food per se, seeds of fiber crops are sources of oils and meals.

TABLE A2 RECENT PERFORMANCE OF EGYPT'S AGRICULTURE SECTOR: Growth in yield levels, 1990–2020.

Commodity/Indicators	Years				Annual growth rate by period				Change from 1990 to 2020
	1990	2000	2010	2020	1990–2000	2000–2010	2010–2020	1990–2020	
Yield	*kg/animal or mt/ha*				*percent*				
Beef/Buffalo meat	134	186	315	319	2.42	5.36	0.19	2.93	139
Sheep/Goat meat	22	29	29	26	4.44	0.20	-1.09	0.43	18
Poultry meat	1.1	1.4	1.4	1.5	2.77	-0.84	0.48	0.63	32
Eggs	71	110	110	109	5.82	-0.61	-0.80	1.82	53
Milk	556	625	909	814	1.72	4.50	-1.16	2.09	46
All cereals	5.37	7.08	7.05	7.07	2.62	0.16	0.11	0.82	31
Maize	5.38	7.47	7.66	7.62	3.19	0.78	-0.06	1.00	42
Rice, milled	4.43	5.93	6.41	5.63	2.61	0.88	-1.14	0.82	27
Wheat	4.99	6.23	6.15	6.52	2.21	0.05	0.63	0.92	31
Fiber crops	0.05	0.06	0.08	0.08	0.97	3.61	3.83	2.96	85
Fruits*	15.33	18.10	21.86	22.44	2.02	1.68	0.48	1.42	46
Nuts	2.87	3.63	5.14	4.39	0.83	2.75	-1.74	2.76	53
Vegetables	22.87	24.45	25.88	24.38	0.53	0.86	-0.44	0.04	7
Oil crops	2.05	2.65	3.48	5.94	2.28	2.74	6.11	3.12	189
Pulses	2.66	2.76	3.11	3.62	0.55	1.48	1.35	0.68	36
Roots and tubers	21.94	23.22	26.34	28.95	0.48	0.98	0.85	1.10	32
Sugar crops	89.86	97.60	87.10	75.18	0.78	-1.04	-1.51	-1.03	-16

Source: Constructed from FAOSTAT data.

Note: Values are 3-year moving averages of t-2 t-1 and t0. Latest data from FAO are for 2019 only, values for 2020 are trend-extrapolation of preceding 5 years of 2015–2019, before the 3-year moving averages were applied to 2020. *Though not food per se, seeds of fiber crops are sources of oils and meals.

TABLE A3 PROJECTED RATES OF YIELD GROWTH, 2020–2050, UNDER BUSINESS-AS-USUAL, WITHOUT CLIMATE CHANGE, INDEXED TO 2020

Crop	2020	2025	2030	2035	2040	2045	2050
				percent/year			
Barley	1.00	1.08	1.16	1.22	1.28	1.33	1.39
Maize	1.00	1.04	1.08	1.13	1.17	1.21	1.26
Rice	1.00	1.01	1.03	1.04	1.05	1.06	1.08
Sorghum	1.00	1.04	1.07	1.10	1.12	1.14	1.15
Wheat	1.00	1.05	1.09	1.12	1.13	1.14	1.14
Other cereals	1.00	1.07	1.14	1.19	1.25	1.29	1.34
Cotton	1.00	1.08	1.16	1.23	1.29	1.33	1.37
Other crops	1.00	1.06	1.10	1.13	1.15	1.15	1.15
Vegetables	1.00	1.07	1.14	1.20	1.24	1.27	1.29
Banana	1.00	1.07	1.12	1.17	1.20	1.22	1.23
Tropical fruit	1.00	1.09	1.16	1.22	1.27	1.29	1.30
Temperate fruit	1.00	1.07	1.13	1.18	1.22	1.24	1.26
Groundnuts	1.00	1.07	1.13	1.17	1.20	1.21	1.22
Soybean	1.00	1.07	1.16	1.19	1.21	1.22	1.22
Sunflower	1.00	1.08	1.18	1.23	1.28	1.32	1.35
Other oilseeds	1.00	1.09	1.19	1.25	1.30	1.34	1.37
Beans	1.00	1.08	1.17	1.23	1.30	1.36	1.42
Chickpeas	1.00	1.07	1.13	1.19	1.24	1.29	1.33
Cowpeas	1.00	1.07	1.15	1.23	1.31	1.40	1.48
Lentils	1.00	1.11	1.23	1.34	1.46	1.57	1.68
Other pulses	1.00	1.08	1.17	1.23	1.30	1.36	1.42
Potato	1.00	1.10	1.20	1.30	1.39	1.47	1.55
Sweet potato	1.00	1.07	1.13	1.19	1.23	1.26	1.28
Other roots	1.00	1.04	1.07	1.10	1.13	1.15	1.16
Sugar beet	1.00	1.08	1.15	1.22	1.28	1.34	1.38
Sugarcane	1.00	1.06	1.11	1.15	1.19	1.21	1.23

Source: IMPACT dataset.

TABLE A4 CHANGES IN PRODUCTIVITY DUE TO CLIMATE CHANGE: Biophysical and economic effects, rest of the world, by 2050

Commodities	Biophysical effects on yields			Combined biophysical and economic effects
	Heat stress	Water stress	Cumulative effects	
	percent change from No CC			
All food crops	-4.91	-1.22	-6.07	-5.24
All cereals	-4.42	-1.07	-5.45	-7.74
Maize	-12.62	-1.88	-14.26	-17.66
Rice	-5.57	-0.11	-5.67	-5.61
Wheat	1.59	-1.25	0.32	0.82
Fruits and vegetables	-3.24	0.12	-3.12	-1.95
Oilseed crops	-4.68	-0.64	-5.29	-6.69
Pulses	-1.19	0.16	-1.04	0.01
Roots and tubers	-6.48	-0.32	-6.78	-4.58
Sugarcane	-6.63	-3.38	-9.79	-10.39

Source: Constructed from IMPACT results.

Note: Values are averages of 3 GCMs: GFDL = General Fluid Dynamics Laboratory; HGEM = Hadley Centre Global Environmental Model; IPSL = Institut Pierre-Simon Laplace; all RCP 8.5 and SSP2.

TABLE A5 EFFECTS OF CLIMATE CHANGE ON FOOD PRICES, BY 2050.

Food commodities	World prices of food		
	No climate change	With climate change	Change
	US$/mt		*percent*
All meat products	3,411	3,581	5.0
Beef/Buffalo meat	4,203	4,317	2.7
Mutton/Goat meat	4,540	4,609	1.5
Poultry meat	2,780	3,024	8.7
Dairy products	588	594	1.1
Eggs	2,539	2,685	5.7
All cereals	268	305	14.0
Maize	212	260	22.8
Rice, milled	427	507	18.7
Wheat	286	293	2.5
Fruits and vegetables	1,184	1,305	10.2
Oilseed crops	552	662	20.0
Pulses	1,132	1,217	7.5
Roots and tubers	448	518	15.8
Sugar/Other sweeteners	412	442	7.4

Source: Constructed from IMPACT results.

Note: Values are averages of 3 GCMs: GFDL = General Fluid Dynamics Laboratory; HGEM = Hadley Centre Global Environmental Model; IPSL = Institut Pierre-Simon Laplace; all RCP 8.5 and SSP2.

B. ADDITIONAL FIGURES AND TABLES: HIGH ADOPTION RATE

Figure A1 and Tables A6–A8 show results for high rates of adoption of climate change adaptation technologies. They complement results in Tables 3.1–3.3 and Figures 3.1 and 3.2 for moderate rates of adoption. High adoption rates increase the effectiveness of adaptation technologies as countermeasures against climate change.

FIGURE A1 IMPACT OF CLIMATE-RESILIENT TECHNOLOGIES ON PRODUCTION OF MAJOR FOOD CROPS, EGYPT, BY 2050 (HIGH ADOPTION RATE)

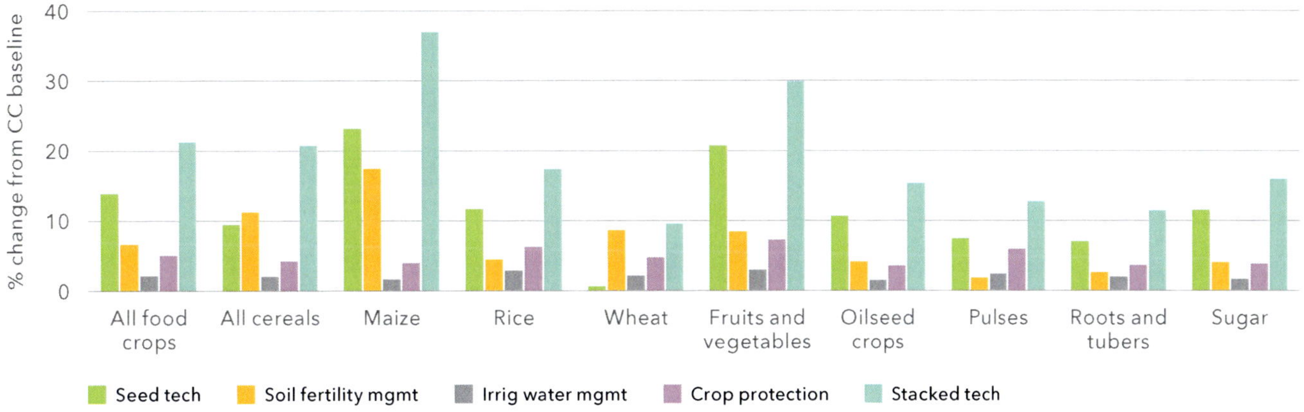

Source: IMPACT simulations.

TABLE A6 YIELD EFFECTS OF SUITE OF CLIMATE-RESILIENT TECHNOLOGIES ON MAJOR FOOD GROUPS, EGYPT, BY 2050 (HIGH ADOPTION RATE)

Food commodities	Yield effects of climate change	Minimum yield change to counter climate change	Suite of climate-resilient technologies				
			Seed technology	Soil fertility management	Irrigation and water management	Crop protection	Stacked technology
	percent change from No CC		*percent change from climate change scenarios*				
All food crops	-6.2	6.6	13.3	6.2	2.1	4.8	19.9
All cereals	-10.4	11.6	9.4	9.6	1.9	4.0	19.1
Maize	-19.5	24.3	17.8	13.3	1.8	3.9	28.8
Rice	-8.5	9.3	9.3	4.6	2.1	4.5	15.0
Wheat	-0.6	0.6	1.6	7.6	2.0	4.3	11.3
Fruits and vegetables	-8.3	9.0	16.3	7.8	2.5	6.0	24.4
Oilseed crops	-12.1	13.7	11.7	6.1	2.0	4.3	18.4
Pulses	-10.0	11.1	7.6	3.8	2.0	4.6	13.1
Roots and tubers	3.6	-3.4	7.3	3.6	1.9	3.6	12.3
Sugar	-13.3	15.3	11.2	5.5	2.0	4.2	17.3

Source: IMPACT simulations.

Notes: Boxed values mean that changes in yields from climate-resilient technologies fully counteract the effects of climate change. Values are averages of 3 GCMs: GFDL = General Fluid Dynamics Laboratory; HGEM = Hadley Centre Global Environmental Model; IPSL = Institut Pierre-Simon Laplace; all RCP 8.5 and SSP2.

TABLE A7 PRODUCTION EFFECTS OF ADOPTION OF SUITE OF CLIMATE-RESILIENT TECHNOLOGIES ON MAJOR FOOD GROUPS, EGYPT, BY 2050 (HIGH ADOPTION RATE)

Food commodities	Production effects of climate change	Minimum production change to counter climate change	Suite of climate-resilient technologies				
			Seed technology	Soil fertility management	Irrigation and water management	Crop protection	Stacked technology
	percent change from No CC		*percent change from climate change scenarios*				
All food crops	-5.7	6.1	14.0	6.7	2.2	5.1	21.3
All cereals	-11.9	13.5	9.5	11.3	2.1	4.3	20.8
Maize	-21.8	27.9	23.2	17.5	1.7	4.0	37.0
Rice	-6.4	6.9	11.7	4.5	2.9	6.3	17.4
Wheat	-2.3	2.3	0.7	8.7	2.2	4.8	9.6
Fruits and vegetables	-9.4	10.3	20.8	8.5	3.0	7.3	29.9
Oilseed crops	-5.7	6.0	10.7	4.2	1.5	3.6	15.4
Pulses	-23.9	31.4	7.5	1.9	2.4	6.0	12.8
Roots and tubers	3.1	-3.0	7.1	2.7	2.0	3.7	11.5
Sugar	-3.5	3.7	11.6	4.1	1.7	3.9	16.0

Source: IMPACT simulations.

Notes: Boxed values mean that changes in yields from climate-resilient technologies fully counteract the effects of climate change. Values are averages of 3 GCMs: GFDL = General Fluid Dynamics Laboratory; HGEM = Hadley Centre Global Environmental Model; IPSL = Institut Pierre-Simon Laplace; all RCP 8.5 and SSP2.

TABLE A8 CHANGES IN NET TRADE STATUS UNDER CLIMATE CHANGE AND ADOPTION OF SUITE OF CLIMATE-RESILIENT TECHNOLOGIES, EGYPT, BY 2050 (HIGH ADOPTION RATE)

Food commodities	No climate change	Climate change	Suite of climate- resilient technologies					
			Average of all technologies	Seed technology	Soil fertility mgt	Irrigation and water mgt	Crop protection	Stacked technology
				'000 mt				
All food crops	-29,139	-27,810	-8,849	-13,025	-20,066	-25,359	-22,220	-4,673
All cereals	-35,957	-32,770	-29,181	-30,531	-30,096	-32,271	-31,731	-27,831
Maize	-17,609	-14,231	-11,348	-12,010	-12,562	-14,062	-13,844	-10,686
Rice	-2,868	-2,770	-2,397	-2,470	-2,654	-2,696	-2,610	-2,324
Wheat	-12,373	-12,558	-12,170	-12,793	-11,636	-12,326	-12,050	-11,547
Fruits and vegetables	12,387	7,777	21,038	18,658	12,242	9,323	11,565	23,418
Oilseed crops	-1,105	-1,025	-917	-936	-994	-1,013	-994	-898
Pulses	-1,653	-1,843	-1,790	-1,804	-1,834	-1,831	-1,812	-1,777
Roots and tubers	4,263	7,244	7,248	6,990	6,472	6,396	6,593	7,507
Sugar	-2,077	-2,076	-1,427	-1,532	-1,881	-1,997	-1,891	-1,322

Source: IMPACT simulations.

Note: Values are averages of 3 GCMs: GFDL = General Fluid Dynamics Laboratory; HGEM = Hadley Centre Global Environmental Model; IPSL = Institut Pierre-Simon Laplace; all RCP 8.5 and SSP2. Negative values mean net imports; positive values mean net exports.

ABOUT THE AUTHORS

Nicostrato Perez is a research fellow, **Claudia Ringler** is the deputy division director, **Timothy S. Thomas** is a research fellow, and **Hagar ElDidi** is a research analyst, all in IFPRI's Environment and Production Technology Division, based in Washington, DC. **Yumna Kassim** is a senior research associate of the Egypt Strategy Support Program in IFPRI's Development Strategy and Governance Division, based in Cairo. **Clemens Breisinger** is a senior research fellow and leader of the Kenya Country Strategy Support Program and the MENA Regional Strategy Support Program, in IFPRI's Development Strategy and Governance Division, based in Nairobi.

Made in United States
Orlando, FL
04 November 2026

90205465R00033